CONTENTS

// WHAT CAN THIS BOOK DO?

Road Trip to Innovation – How I came to understand Future Thinking is a book about understanding ways of approaching and building innovation. The road trip itself is part of the narrative of the book as I travel from city to city, interviewing people in order to find out, together with you dear reader, the following: what does innovation encompass, how to spot trends, how to use trends for creating innovative products, services and strategies, how companies approach innovation in terms of method (e.g.: design thinking, trend consulting) and how are trends transformed in a strategic asset inside companies (e.g.: strategic foresight, strategic planning, forecasting).
The book will offer an introduction into the field of future studies and targets to raise awareness of the innovation and trend research industry. It will explain complicated notions in easy language and with a friendly approach. *Road Trip to Innovation – How I came to understand Future Thinking* also has educational purposes for students and trainees; it offers a holistic view on methods and tools for building innovation to young managers setting up innovation strategy in their companies and furthermore, it is a good introduction to the trends and innovations industry for anybody who is interested in the topic.
My findings are based on literature review and data drawn from interviews and case studies of companies in Europe and the United States. The interviews were conducted with companies that encompass a broad spectrum of approaches and with people of varied backgrounds, as I believe in the power of poly-social groups. Why all these companies and why these people? There is no financial reason. They were either found by natural research (a.k.a. appeared first in the Google search engine) or recommended by Nils Müller, as he is the engine behind this book, mentoring me and connecting me with people from the industry. There is no advertising reason; they are just some examples out of several interesting companies that are out there.

The Future Navigator that stands for all the contents of this book with methods, tools and applications of the innovation process was created and inserted in this book at the end. It started as a plain piece of paper with two arrows that cross each other. It was gradually filled with the approaches presented in this book in order to start a conversation regarding their meaning and purpose. It can be looked at as a chess table. Anybody who has additional information and strong motivations can modify it.
Road Trip to Innovation – How I came to understand Future Thinking is the fruit of my thinking on all the aforementioned subjects: It is based on the literature that I've read and the fascinating and intelligent people that I've spoken with. As we are all different, special and unique persons, anyone could write this book in their own way and interpret things according to their own dimensions. This is a first published wave from the sea of opinions.

VISUAL ELEMENTS

Visualization is very important when it comes to understanding complex things; therefore, every chapter is sketched by Anna Luise Sulimma who invented a visualization method for the trend industry called trendsketching. You can find her on TrendSketcher.de.

Here is the legend of the visual elements we are using in the following illustrations, mind maps and icons.

evolving trend

weak signal

emerging issue

observation
environmental scanning

interpretation

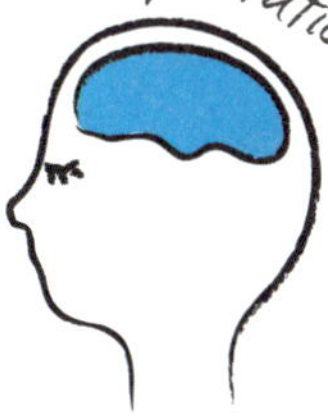

design thinking

an orientation instrument

engineering solution

ALL ON BOARD?

The Basics

Tensta Konsthall by frontdesign.se

I AM A WALLFLOWER

I didn't quite realize it until my Coolhunting instructor at Instituto Europeo di Design Barcelona and dear friend, Christina Bifano, pointed it out to me. It's true; I am that shy soul sitting in the corner of the ballroom, waiting for somebody to invite me for a dance. However, being a wallflower has certain advantages. From where I stand I can observe closely the goings on around me and ponder the behaviors and appearances of others. I am naturally curious. I want to know *why* and *what* lies behind the obvious exteriors. Don't you?

People ask me:

> So what do you do for a living?
> I am a trend researcher.
> Excuse me? Can you repeat that?!?!?
> Ok, I am a trend researcher working in the field of innovation.
> A what – working in the field of what? I don't understand: What is it exactly that you do?

This conversation is probably very familiar to people working in the field of trend research and to students or professionals from the fields of innovation. Meeting for a beer and talking about professions has a whole new dimension. We are like a new breed of civilization: we ourselves and people we come in contact with are constantly amazed with the existence of our job description. Always on, always wondering about what comes next, we can't escape, not even for a second, our future thinking. While I was studying coolhunting at IED Barcelona, I often ran into difficulties trying to explain to others what I was studying. This is one of the reasons why this book was born.

Long story short, after my studies I had the opportune chance to meet Nils Müller, CEO and Founder of TrendONE Germany. It was a rainy September day in Berlin. I was looking for a job in trend research and funnily enough, my previous boss from Lowe PR Bucharest, Hortensia Nastase, had just seen a TrendONE keynote speech and said: "These guys are great! You should contact them". Luckily, they accepted my invitation to meet at their office in Berlin.

A few hours before the meeting, TrendONE called me and asked if I could come to The Ritz instead. Yeah, right, I *always* go to the Ritz. I was just testing the waters seeing if this could lead to a possible job interview and suddenly the whole meeting took on a whole new meaning for me. It's always about contexts, isn't it? And the context here was, well...The Ritz. After some minutes of racking my brain, figuring out what to wear, I decided to play it cool. Classic black with a handmade necklace that I bought at a small craft fair in Bucharest. I entered the hotel, politely greeted the bellhop and took a seat in the extravagant lobby. A few minutes later, Nils Müller made his appearance wearing an impeccable suit, a chic scarf and pulling a small metallic luggage case. He says sorry for being late; he was 2 minutes late! As I was coming from Spain, where I lived for one year, being late had a totally different meaning for me. There, they haven't decided yet if showing up a half hour after the proposed meeting time is late or not. In an attempt to lighten up the atmosphere I comment about Spanish people and punctuality and all I get is a straight answer: "Oh, no, we don't do that here". Ok, noted.

From there we went to eat at Vapiano, an Italian restaurant nearby where pasta is cooked in front of you in an open kitchen. I was nervous so I couldn't eat. I just ordered a Coke. Since Nils was eating, I had to do the talking first. Among other things, I said: "You know, while I was studying coolhunting I couldn't find a book that explained what the trends and innovation industry is all about, like a holistic approach to methods and processes. There are hundreds of terminologies out there that confuse me. I wonder how clients can deal with this." With time, I learned how proactive and full of brilliant spontaneous ideas Nils is, but back then, I was surprised by his answer: "Ok, I got you. Then why don't we work together on this project? Why don't you write it?" A few minutes later I went out of the restaurant with the sensation that my life was about to change completely. I was going to write a book.

// WHAT IS INNOVATION?

[ĭn'ə-vā'shən]

1 THE INTRODUCTION OF SOMETHING NEW;

2 A NEW IDEA, METHOD, OR DEVICE: NOVELTY.

Innovation is a change; it is "new stuff that is made useful" as Max McKeon writes in his book *The Truth About Innovation*. Therefore, the innovation process, in this book, refers to the journey one has to make in order to obtain a new, cool, innovative idea and how to manage it. In a nutshell, it is about how to get the idea and what you can do with it to make it valuable.

The Organization for Economic Cooperation and Development (OECD) report on the *New Nature of Innovation* states that a new nature of innovation is emerging. How? "Innovation is no longer mainly about science and technology. Firms can innovate in other ways. Co-creation, user involvement, environmental and societal challenges increasingly drive innovation today. Collaborative, global networking and new public private partnerships are becoming crucial elements in companies' innovation process."

In the early 1990s innovation equaled product development; this approach was actually much too narrow. But the modern approach is also problematic because the plasticity of this term turned the meaning of innovation into a catch phrase.

Nowadays, everything that is providing value for a company is being referred to as an innovation. For some companies it may be as simple as gathering around for a pizza on a Friday night at the office in order to create a better organizational climate thus internally providing a better basis for innovation and creativity. As for others, it cannot be called innovation until the brilliant idea that provides value is marketed – like creating an app that tells you the ingredients of the pizza just by scanning the barcode of the box. So, it all depends on how deep you go into the term of innovation, if just about anything can be interpreted as an innovation. But it shouldn't be this way. In my opinion, the prior example dilutes its powers.

THE FIRST AIRPLANE

These days it seems that innovation as a noun has been thrown into so many contexts that it has become hype. This makes it difficult for its meaning to be securely locked in a box. For this reason, it becomes subjective and highly interpretative. The good thing is the popularity of the word can only do well to the world. Innovation turns out to be the source and, in the same time, the target for more and more strategies. Martin Kruse, futurist at the Copenhagen Institute for Future Studies (CIFS), states that "research into innovation offers insights that can benefit everyone working with organizational development, management, finance or product development."

Defining what innovation means was giving me a hard time. Every answer I got was competing in being more poetic than the one before. Here is one particularly poetic definition coming from the *Future Driven Innovation* paper from CIFS: "Shakespeare calls Man the 'paragon of animals,' the greatest work in the history of evolution. This is because we have a brain with 100 billion cells and so the ability to change the world around us to give ourselves a better chance of survival. We develop ourselves through creative production, a process that, since the dawn of financial markets, has been known as innovation."

From the same source, here comes the practical definition: "Innovation has historically been regarded as inventing a new product, producing it and putting it on the market. It is the result of a creative process with emphasis on value creation. Today, innovation can happen anywhere in the company. The value created can be internal, as is the case with human resources, or directed out to the customer as a product on the shelf. The point is that value is created and reaches the customer in some way."

What is important to understand from the start is that there is a difference between innovation and having a creative idea. Both processes can be innovative in their own way but we shouldn't confuse them. Kruse told me a little story that helps to keep in mind the difference between the two processes.

The first airplane was actually visualized by Leonardo da Vinci in the 1480s. That was just a creative idea. The idea became an invention when the Wright brothers made a plane out of it, which lifted from the ground in 1903. But the airplane, at that point, was still not an innovation because it was not yet marketed.

Twenty-one years later, Continental Dusters (which subsequently became part of Delta Airlines) gave the airplane its first commercial use, dusting crops. This is the point in history when the airplane became an innovation.

Similar points of view to Kruse's are those of Richard Florida and Martin Kenney, in their book from 1990, *The Breakthrough Illusion: Corporate America's failure to move from innovation to mass production*, consider invention as a breakthrough and innovation as an actualization.

// THE FUTURE THINKING MINDSET

Magnus Lindkvist

What drives somebody to become a trendspotter? How does somebody arrive at such a profession? Some people have a clear picture and follow a more or less standard trajectory towards this title while others fall into it seemingly just by happy coincidence. Magnus Lindkvist belongs to the latter category and he has one of the most interesting stories I have heard so far.

He is an expert on present and future trends, a futurologist and lecturer focusing on business intelligence and trends. Before founding his company, Pattern Recognition, he worked as a brand strategist and then as a planner and management group member at the brand strategy agency Grow. He is the author of the book *Everything We Know is Wrong: The Trendspotter's Handbook* and was voted Sweden's Business Speaker of the Year in 2009. Lindkvist is also an active member of TED, a hub of brilliant minds based in California.

It was one of those mornings when nothing seemed to work. I didn't hear my alarm clock, missed the bus and when I finally arrived at the office I noticed that my laptop cable was missing and I was running very low on battery (somebody at my office switched it with his own and took it home!) I arranged with Lindkvist a telephone interview. As I didn't want to miss anything of what he was about to tell me, I planned to record it with my computer's software – hoping my battery would hang on in there for the duration. And it did. (Thanks Lenovo!) It was a fascinating 45 minute conversation.

THE BEGINNINGS

"It was just one of those fortunate accidents", Lindkvist says, regarding Pattern Recognition, which was founded in 2005. Back then trend spotting was his hobby and he had been giving trend lectures for free, for whoever wanted to listen. He has always been interested in storytelling and he even wanted to make movies but ended up *just* writing a novel which was also just for fun. "I always wanted to form a narrative, to try to describe events and what is happening", says Lindkvist.

Prior to August 2005, Lindkvist had been working for advertising companies and he was quite miserable so he decided to change things: "I was booked to fly with my boss to a meeting in Finland – I woke up in the middle of the night and I felt I'm not going to do this anymore. That's when I decided to found my own company".

He was asked to write a trend newspaper for a media company a few months earlier and he suggested the name "pattern recognition" but they didn't like it, they said it sounded too academic. So, Lindkvist took the name and started his own company. "I had no idea what I wanted to do. I felt trends were somewhat interesting". A former boss of his had said that one can't have a business by giving speeches, it just doesn't work. So, Lindkvist said, "Ok, I'll see what I can do". He didn't see an actual need for this kind of service; instead he tried to explore what the needs were. He wanted to find a model.

He started by selling reports to advertising agencies and then moved on to doing full day seminars for companies but then again, he says, "corporate meetings are boring. I found out that I really enjoyed the 45 minute keynotes speech when you can really deliver a flash of energy like an energy drink, like a Red Bull. So I gave many, many bad speeches. It was painful but I felt like somehow they were developing me as a speaker. Since then it was a long positive spiral. My speeches got better and better". In fact his speeches got so good that he is now writing his second book and is traveling around the world giving his power-infused keynote presentations.

SNEAKING IN

But still, where does this inspiration come from? I asked myself, "but where does he look for it?" Like anybody else, he reads books and magazines, hears podcasts. That's what we all do but each of us sees things differently. "What I search for is the idea which will resonate with people. Showing somebody an example and saying look at this gadget, this machine, this Chinese artist won't necessarily resonate. What will resonate will become a virus. You adapt yourself to the immune system of the host so you can sneak in. You take something twisted and make it familiar or you make something familiar twisted," says Lindkvist.

What about companies and brands? Can it be that brands fail because they are scared to take risks and see themselves as being too 'traditional' to innovate?

For example, in the case of the Greek crisis he always brings into speeches the *Uncertainty Avoidance Index* of Professor Geert Hofstede, which measures different countries' relationship to risk taking, familiarity with the unexpected and comfort in trying new things. Greece is one of the worst countries in the world when it comes to this topic. They have one of the highest Uncertainty Avoidance Index. Could the fact that they seem to hate trying new things possibly be responsible for the country's crisis?

Lindkvist makes these kinds of connections: "Many companies are really bad at trying out new things. So, my job is to take the information that we all have access to and decode it. I want to leave people with some hooks in their brain", he says in his consummate energetic voice.

BETTER FUTURE THINKING

Lindkvist says he works for better future thinking. I was quite curious what future thinking referred to. I have to admit, it sounded quite provocative. I found out that Daniel Gilbert, in *Stumbling on Happiness* does a great job in describing how we think about the future in the wrong way. We tend to be overconfident or too vague or oversize things in the present. Lindkvist was heavily inspired by Gilbert's book. In terms of what better future thinking means, for Lindkvist it is about being more imaginative and tangible as it inspires other futures: "If you ask people about the future these days they seem to be very pessimistic: the recession, climate change etc. You have to look beyond the present because we all know the recession will end, we know it's all going to pass. We know climate change is about an innovation or two away from making CO_2 redundant. Sweden has cut the carbon emissions in half since the 1970s so it can be done and quite easily. So, let's look beyond today and really start using the imagination for what is possible tomorrow. That's one part of better future thinking."

When people think about the future they have some random thoughts and then, instead of sharing their ideas and finding ideas that can move others, they just stop and don't think about it anymore. "You should ask yourself: what is the future scenario which will inspire your employees to innovate, your colleagues to do better things and your company to function in better ways. These are the main ingredients of better future thinking and this is what drives me to go out there and speak, give inspiration and energy," says Lindkvist.
"Will the need be there for a PowerPoint-making speaker or a trend research agency with scouts around the world in the future? Of course not! We will have a completely new tool for it. That could be an algorithm crunching all data in a millisecond that replaces the scouts. It could be something that we haven't seen yet," adds Lindkvist.

THE 78 YEAR OLD AND THE PINNACLE OF INNOVATION

"Bin Gordon said that if you want to be innovative don't hire 35 year olds. I found that to be a very interesting quote. What it means is that the 35 year olds have a young family, a big mortgage and don't want to take risks. [Lindkvist starts empathizing with the thought]: I am 36 so I completely relate to that. We want to earn easy money; we don't want to try new things. The 18 year olds are obviously full of risk taking but they become 35 and will still face that problem. The 78 year olds paid off their mortgage, have failed several times in life, and are not necessarily scared of it. They would be magnificent innovators and pioneers. That is a thought".
Of course, there has been said for quite some time now that people need to work longer; we can't retire at age 40. Even the system where people retire at age 65 is unsustainable. But we need to give retired people something to do. "What if the peak of people's imagination and innovation comes at age 78? Not at age 35. Look at how many companies were founded and how many insights were come across when people turned 60 and older. This really turns society and the work place upside down. It changes how we work and how society functions, and quite frankly, it gives me hope for the future", says Lindkvist.

// *MAKING SENSE OF IT ALL*

With all this being said, the innovation industry is still somehow at its beginnings and it lacks structure. A lot of terms are overlapping. I started asking myself, "What is the difference between all the cool terminologies that describe innovation? What is the role of trends in innovation management? What are the research methods? What does strategic foresight or future insight, future thinking or strategic planning and forecasting mean? Does the term future management relate to all of them? And how to decide upon the relevancy of the information?" All these questions made my head spin. This book tries to sort things out and reduce the complexity of the processes to a brief and understandable framework.

My encounters show that trends, taking different shapes, are the basis for most innovation thinking process. Trend research is where it all starts when it comes to the future. I am not talking here about "red is the new green" but about trends that shape social, cultural, economic, technological and political behaviors.

> "In a world that is constantly changing, there is no one subject or set of subjects that will serve you for the foreseeable future, let alone for the rest of your life. The most important skill to acquire now is learning how to learn.
> / John Naisbitt"

Nils Müller always compares the fragmentation of the innovation industry to that of the consultancy world in the 1960s, when the concept of consultancy was at the beginning of its existence. The funny thing now, when we look back, is that the term and profession had a generic meaning. Consultancy firms offered their services in any domain. With time, the industry developed and different types of consultancy started to appear. Now we have consultancy for legal, financial, IT and so on. The trend research and innovation industry, if we can call it an industry yet, offers a forward thinking type of service and it is as known and structured as consultancy was in the 1960s. The lack of structure is certainly another reason why this book has to be written.

We are living in a new media society. A decade ago this book couldn't be published for several reasons: a) I was still in school. b) you couldn't publish a book unless you had a doctoral thesis, some distinction or award. c) it was much too expensive. Now, even if we live in the digital world we print more books than ever. Of course that doesn't mean that everybody can do it right. While Condé Nast closed down over 10 magazines in the last 2 years, cutting edge magazines such as Metal, Apartamento and Monocle are thriving. Because they offer (and I am paraphrasing their own words here) cutting edge content and they refuse everything that is ordinary. Why do they work? Because they are offering something innovative.

Browsing the Internet to find new insights, I stumbled upon a citation that suddenly revealed how things stand in the new media society: "Growing up assuming you can publish whatever you want to say, to whomever you want to say it, is going to make people different". Great quote, slightly paraphrased, from an article Russell Davies wrote about Clay Shirky's book *Here Comes Everybody.*

I am on a difficult mission, to find out, understand and put in order the concepts from a huge heap of information. Our road trip starts with a few basic theories delivered by some smart people. It is necessary to grasp the meaning of these theories so we can hop onto the next train.

// TRENDS

"No matter whether we speak of open-plan kitchens or technological innovation, the innovation is a response to a trend", says CIFS' paper on Future Driven Innovation. This statement is completely relevant for our journey here – the fact that trends and innovation are interdependent. Over the past years, the word 'trend' has been associated with many things in different contexts. The most popular situation in which trend is 'at home' is fashion. When people hear the word trend they already start thinking of the next season's style. Actually, a trend is about more than this. Fashion is only a short-term look into what is going to happen in the future. Clearly, a trend means different things for different people. But most of the time it is associated with something cool. A trend is something cool. Even saying the word trend is cool. You are a trend researcher? Wow! Cool! See what I mean?

But what is cool? Malcolm Gladwell said in 1997 in his article *The Coolhunt* from The New Yorker that "there is no coherent philosophy of cool". What we know for sure is that cool sells products and services and companies are already aware of the immense power of it. Cool is ever changing; it evolves according to its place in time and space. It can receive different names; therefore there is no definition of cool aside from something funny, nice, and different. I will not spend much time on explaining what cool means as this is the subject of already too many books. I think we are all aware of its meaning, both colloquially and commercially. If the desirability of cool is so widely felt, businesses need to continue to track what cool means right here, right now and tomorrow as well.

Cool is tracked and spotted in trends. People have the tendency to ask "what's the current trend"? A proper answer to this question stands in the fact that trends are declined in various forms: society, culture, area, time and period. Also, trends vary in terms of intensity and duration. So we have to deal with micro, macro or megatrends. There is no such thing as THE trend. **Trends come in different sizes!** The growth of a trend is also in direct connection with the area where it is present and the number of people it affects. The more we go from micro to macro or mega, the more the trend grows from a concrete example to one that covers a wider distance and affects a larger amount of the population. Don't panic! Clarifications and examples will follow in the next chapters as we continue our journey.

Henrik Vejlgaard says in his book *Anatomy of a Trend* that a trend sociologist will talk about a trend moving from the trendsetters to mainstream and in this case, trend refers to a process of change. In popular magazines we see headlines like New Furniture Trends. In this case, trend means product news. Someone who is working in design can talk about the trend in car manufacturing. In this case, trend will mean product development.

All trends are categorized here by their size in reference to time span. I will just give you a quick example of what I mean. A megatrend is definitely a global trend as it touches many people. **Megatrends last decades, they affect many different aspects of society, and they involve a complex process that often includes politics, economy and technology.** 'Healthstyle' is something that matches the description and it is one of the megatrends presented in the *Trendbook 2012* published by TrendONE. Health is increasingly becoming a consumer item and used for self-expression. A whole new life-philosophy is built around this core which consists of health awareness, body, spirit and soul and which promotes prevention as opposed to intervention. Foreseeing is the name of the game. Health no longer means the absence of illness, as it is increasingly becoming a consumer good, a form of self-perception and a source of strength. 'Healthstyle' will have a dramatic impact on the consumer and service markets, as well as on the forces at play in the entire health industry. Consuming organic products is a sub trend of 'healthstyle' and namely a macrotrend. **A macrotrend has a shorter time span of 2 to 5 years** and it can possibly grow into a megatrend if it starts affecting more people, covering a bigger area.

A microtrend lasts from 6 months to 2 years. In relation to 'Healthstyle', a microtrend would be manufacturers creating intuitive technologies such as the iPhone app 'GoodGuide', which helps shoppers to find safe, healthy and green products. **Trend research** and analysis involves observing (spotting) and synthesizing numerous small signs, to determine if they follow any patterns or if they have linking characteristics. Observing people, monitoring a wide range of media and sorting information into what will be announced as a trend are some of the methods used, according to Henrik Vejlgaard in *Anatomy of a Trend.*

Why do we need all this? Because spotting trends can give us a better understanding on what is going on around us. Slowly, this became a marketing tool. Eric Pfanner's 2006 article in the New York Times entitled *Agencies Look Beyond Focus Groups to Spot Trends*, states: "Trend spotting has grown in importance as marketers have become frustrated with some of the traditional tools of market research". **Trend spotting** is about getting advanced knowledge of what will happen in the market, thus a useful tool for creating a competitive edge in businesses. Most new product developments take a considerable amount of time (in the car industry up to 8 years), and the better you adapt to the market in a short period of time the more successful you are likely to be. **If a business can count on a new trend to grow, it can align its product development with that trend.** "As more and more industries become aware of trends, there is a growing need to understand trends better and there are several kinds of experts with a variety of functions and roles who are ready to give advice", says Vejlgaard.

// FADS AND HYPES

We must distinguish trends from fads and hypes. Ultimately all fads have the same characteristics: they are very short stories that revolve around some new, innovative products. But they never go further than this.

According to Wikipedia, a fad is said to 'catch on' when the number of people adopting it begins to increase rapidly. The behavior will normally fade quickly once the perception of novelty is gone. Very often, fashion is associated with fads. If we come to think of it, besides the little black dress, a T-shirt and jeans there are not so many constants in fashion.

Distinguishing between trends and fads might be difficult because fads are much more visible. This is why you have to go one step backwards so fads don't block your view. Naisbitt mentions an interesting aspect about the fact that the fads of one trend can change, while the core information of the trend doesn't. **What we do is a trend and how we do it is a fad.** Take for example the concern for our body shape. It has been there for years but how we take care of it has changed from just jogging to complex fitness & diet programs, yoga and massages.

Richard Watson, author and publisher of *What's Next* future blog, said in an interview given to Giselle Weiss for the Global Investor in April 2010 that "a fad is a pattern of attitudes or a behavior that is short lived or only involves a small number of people. A trend is a pattern of attitudes or behaviors, too, but it has more rigor and resilience. It could last decades as opposed to six months or a year. And trends are larger".

Fads are embedded into trends and are manifestations of trends.
/ John Naisbitt in *Mind Set*!

The text book definition of hype is similar to a fad as it refers to the over enthusiasm and subsequent disappointment that typically happens with the introduction of new technologies. Now, hype doesn't refer only to stuff related to technology but to anything. Seems like, the innovation and trend research industry borrowed terms from the 'scientific' fields instead of inventing new ones and had started using them with new meanings. Probably this is one reason why my head spins when I read all these terms that are familiar to me in other contexts.

"Yes, this makes me crazy. For me, hype is the buzz created around anything, anyone or any circumstance", says Christina Bifano. "Remember that song Don't Believe The Hype by Public Enemy? They're not talking about a new technology". Wait a second, does that mean that the hype is a meme?

// THE MEME

Do we look smarter if we use complicated words that nobody understands? Hmm...

This concept is so simple and therefore so attractive to people. Roughly, the meme is about imitating behaviors. The definition of a meme according to Wikipedia is "a unit of cultural ideas, symbols or practices, which can be transmitted from one mind to another through writing, speech, gestures, rituals or other imitable phenomena."

Princeton University's online dictionary refers to a meme as being "a cultural unit (an idea or value or pattern of behavior) that is passed from one person to another by non-genetic means (as by imitation); memes are the cultural counterpart of genes".

According to Oxford dictionary online, "When Richard Dawkins coined the word meme in his 1976 book *The Selfish Gene*, he wanted a word like gene that conveyed the way in which ideas and behaviour spread within society by non-genetic means. Since then the word has been picked up to describe a piece of information spread by email or via blogs and social networking sites."

To put it in simpler words, memetics is the discipline that studies how memes or elements of cultures and trends evolve and spread between people over time. "Memes are the basic building blocks of our minds and culture, in the same way that genes are the basic building blocks of biological life", says a report from InSites Consulting, *Forward Thinking series on Teen-memes.*

A meme can be a simple idea, or a complex idea. The most important point about a meme is that it propagates. Memes are like gold for brands. They spread like epidemics especially among young people. Language memes are sayings or phrases that are typical, like the word 'Ok' together with its hand sign. These both could be considered memes due to their wide spread, international use and street origin. Fashion memes are the items youths wear to express their belonging to a certain group, their haircut, their type of make-up, piercings, etc.

But how do memes apply to brands? Some people call it memetic branding. As memes are things that spread fast and uncontrollable, brands wish to actually become a meme in itself. "Looking at teen-memes, [says the same report from InSites Consulting], is valuable for brands for a number of reasons. First, brands can act as references, personalities or icons. On the other hand, memes invade our minds, change perceptions, attitudes as well as behavioral patterns and get passed on by imitation. Memes are therefore excellent communication vehicles".

The best way to understand all this is looking at youth, as they are the most exposed and active group of people. They create their identity by collecting and clustering an individual set of memes. If the difference between the memes they adopted and certain mainstream behaviors are very different than those people who are 'memeing', they will be perceived as being part of a subculture.

There are still more articles based on the theory of memetics, but I decided to continue the road trip now with the acknowledgement that **the meme is about imitating simple behaviors which relate more to language and fashion than to perceptions, emotional states and experiences in general.**

// THE TIPPING POINT

Malcolm Gladwell developed the tipping point theory and wrote the book with the same name aiming to explain the emergence of trends. He advises that the best way to understand trends is to think of them as epidemics. When some people think of shoes, rubber may not be the first material they associate with comfort or style but this didn't stop Havaianas from spreading like a virus. They started as the first inexpensive cloth shoe made with Brazilian coffee farmers in mind in the year 1907 and then transformed into a very affordable rubber sandal available anywhere from gas stations to specialty stores in Brazil. Havaianas have become, in the past 8 years, a fashion necessity all over the world. Outside of Brazil, fashion conscious New Yorkers started wearing them in the 1990s, shortly thereafter Argentinean Rolinga subcultures adapted them as their own and well traveled Europeans were quick to follow. A brand of national pride, Havaianas are still *infecting* people today with new styles and marketing strategies.

At the center of the tipping point theory there is the possibility of sudden change. **The tipping point is the moment of critical mass, the boiling point.** Gladwell continues to describe the phenomena in the book by setting up some rules called "laws of the tipping point" and I advise you to read it. There is not enough space here to tell the whole story and it is a *must* for the trends industry.

Ideas, products, messages and behaviors spread just like viruses do.

What you have to keep in mind now is that the tipping point of any product or service might come (it's not always happening) and this is the point when producers lose control of the sales, masses start using the product.

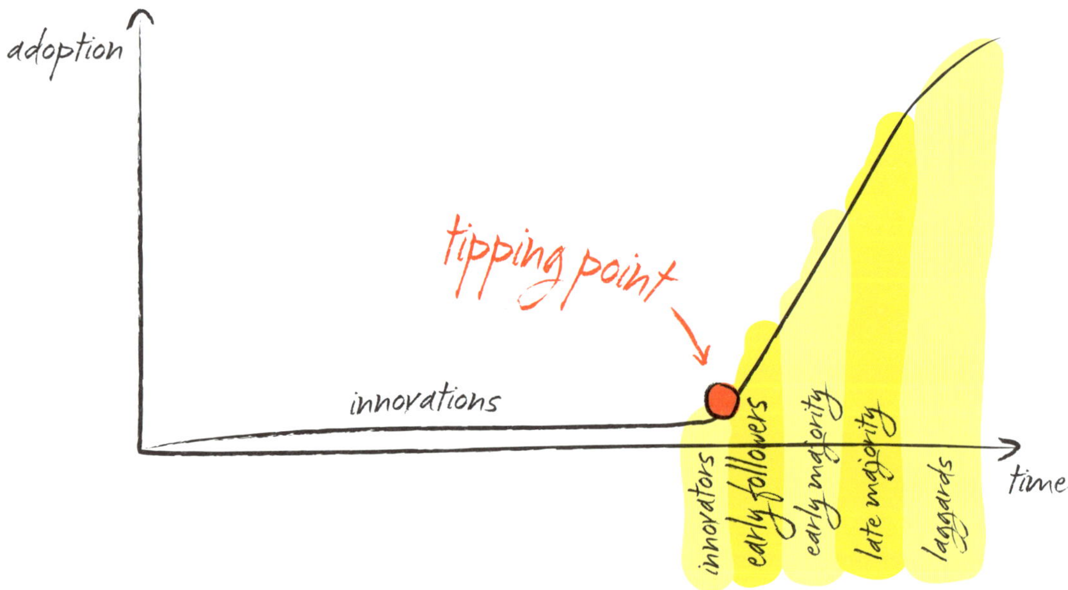

// THE DIFFUSION THEORY

Diffusion is very well linked with the tipping point theory. You'll see why. Academics call the process by which new ideas or products are accepted by groups of people, The Diffusion of Innovation. The term is coined in 1962 by Everett Rogers, professor at University of Mexico. He set up five steps of the diffusion process and he also tried to build some criteria by which innovations spread out faster. He explores how the social environment influences the way a new innovation spreads. This is what it looks like:

STEP NO 1 THE INNOVATORS

2.5% of the users. What motivates them to try out an innovation? They simply become interested and curious in the idea.

STEP NO 2 THE EARLY ADOPTERS OR THE FOLLOWERS

The next 13.5% of the users. Are integrated in the local society, are leaders of opinion in their social surrounding, they are respected and serve as role models.

STEP NO 3 THE EARLY MAJORITY

The next 34% of the users. They are the close friends of the early adopters, have their own opinions and think before adopting a new idea.

STEP NO 4 THE LATE MAJORITY

The next 34% of the users. In contrast, they adopt innovations only because of peer pressure and economic necessity. They are skeptical and cautious.

STEP NO 5 THE LAGGARDS

The last 16% of the users. They don't have any opinion.
They are isolated, have the past as a point of reference and are suspicious about the 'new.'

So, once innovation occurs, it may be spread from the innovator to other individuals and groups. Adoption and diffusion of a new innovation follows an S curve, like the one from the tipping point theory. After a slow initial acceptance by early adopters, adoption reaches a tipping point where usage explodes.

I read about Everett Rogers's theory in the book called *Coolhunting. Chasing Down the Next Big Thing* by Peter Gloor & Scott Cooper. It is funny how they tell the story there so I will paraphrase them: a lonely genius guy creates the innovation. But as he is lonely, it takes some time until other people find out about the innovation and convert it into a trend. Trendsetters (see definiton below) come into the game and help the trend spread using their social skills. Pretty soon, adopters want to jump on the bandwagon and take part in this. Next, they release the trend into the outside world, to the people who will carry it to the tipping point. This phase starts with the early adopters again. That's when the innovation starts to *tip*, grow in popularity very fast.

Peter A. Gloor's COIN theory (Collaborative Innovation Network), presented in the aforementioned book of Gloor & Cooper is also interesting. Gloor is a Research Scientist at the Center for Collective Intelligence at MIT's Sloan School of Management and his theory refers to groups of people that act as a beehive, likeminded people that produce winning ideas from a coolhunted idea, by sharing information and work. They say that, if we want to find cool trends, we have to look for "COINs working together in swarm creativity." **The trendsetters** are here referred to as the 'COIN'. Trendsetters are the ones who convince others that it's time to adopt an innovation. They are the ones smoothing the path from step no. 1 to step no. 2 of the diffusion theory. Trendsetters are potential-spotters; they can identify cool things in an incipient form. **They are those people with a vision.**

Trendsetters are very busy people. "They focus on styles and tastes that are completely new and have never been seen before, [that are outside of mainstream], that are in continuous style development, not overhyped by the media as a trend", as Vejlgaard mentions in *Anatomy of a Trend*.

They not only identify the latest cool things but also tell the story further by being an example to the others. As soon as the others get it, the trendsetter has to go out there and search for more. Because when something hits mass market, that something is no longer cool for them. What follows? In most of the cases the change is sudden and 180 degrees wide. If we think of it, almost all big cultural movements have a counter movement. Every action has a reaction. See the curvilinear art nouveau style from the turn of the 20th century being followed by the geometric art deco movement in the early 1920s.

Cool people disseminate cool trends. They are looked at like carriers of innovations. We will notice during our road trip that opinions here differ in relation to the source of innovation. Some look at trendsetters (which is mainly related to the coolhunting term), some of them look at technology and innovations (trend research), some at products and consumers (consumer insights and trends).

Malcolm Gladwell argues in *Consumer Tribes* by B. Cova, R. Kozinets, A. Shankar that "for some time now and in the youth market in particular, insight, innovation and knowledge has trickled up from the street, with all the major players in this market wanting a window on the world of the street. The world of coolhunters meant global companies paying for the transmitted thoughts of kids on the street. Not just any kids, but those considered to have the cool touch".

So, are trend researchers and coolhunters both looking for changes? What is the difference between them? The timeframe. Coolhunters look at things that will happen within a small timeframe, a nearby time horizon.

The term **coolhunting** is often used in the fashion industry. It is actually not only related to fashion but more with investigating and analyzing people and their behaviors in public. Coolhunting is searching for trends by observing people, media and society in general and spotting the new things that will become cool. The difference between coolhunting and trend research is not very well defined and opinions differ. Some of the people that are trend researchers have never heard of coolhunting while some university master's titles bear the name coolhunting and they might include classes related to trend research.

If the time perspective of the research is longer and includes aspects of life other than fashion, trend researchers will look at **lead users.** The lead users are people who are enthusiastic about using the newest technological gadgets and who continuously want to (re)invent things. 'Lead user' as a term was coined by Eric von Hippel in 1986. According to his definition, lead users face needs that will be general in a marketplace – but face them months or years before the bulk of that marketplace encounters them, and are positioned to benefit significantly by obtaining a solution to those needs.

Something that is going to affect a lot of people has to be seen as part of a process. It will not just pop up out of the blue. This is why identifying behaviors, even patterns maybe, new cultural behaviors and so on is the key to future development. **This is why it so important to know what is going on now!** Being constantly updated with trends you can get a hint of what is going to change in the future, where the market is heading. Also, it is important to know that not all trends will be successful. Some of them will be just fads, some of them not even so and others will become big movements.

All these elements that I talked about so far represent basic theories and thoughts on how trends are created. Memes, contagiousness, diffusion, tipping point, lead users, trendsetters, and COINs guide us to a better understanding of why human behaviors take one turn or another and in some cases why they take the same turns. Now that this is clear, we can ask ourselves what is the connection between innovation and trends and start training our minds to be more consciously aware of what surrounds us and what's to come. So, let's hit the road! First stop: Helsinki.

WHERE DO IDEAS COME FROM?

Scanning

WEAK SIGNALS

Elina Hiltunen

HOW TO BOIL A FROG

If you drop a frog into a kettle of boiling water it will definitely try to escape. The reason being the temperature difference is too extreme (20°-100°). However, if you put a frog in a kettle of water at room temperature and then start to slowly warm the water, the frog will hardly even notice when the water starts to boil.

The moral of this story: we easily adapt to changes; we may not even notice them if they happen gradually. With this funny and insightful tale I began my morning discussion with Elina Hiltunen in Helsinki.

Elina Hiltunen, founder of What's Next Consulting, carried out her doctoral thesis about using weak signals in organisational environment and has been interested in this subject since 1998. She is now Finland's leading expert on weak signals and focuses on issues like anticipating and innovating future changes by utilising them. Through weak signals (the first bubbles in the water as it heats) one can foresee changes.

Weak Signals are signals of emerging issues.

While talking to Hiltunen, I started to understand what this weak signal thing is all about and how they are useful futures tools for companies. Elina offers companies methods, education and inspiring lectures about weak signals, which she briefly defines as signals of emerging issues. In practice, she explains, "they can be news stories or observations about technological and social innovations, posts in the social media, observations of novelty products in exhibitions, or simply a modest wall-sticker of an alternative movement".

Igor Ansoff, a Russian American applied mathematician and business manager who is known as the father of strategic management, coined the term 'weak signals' in 1975 and, according to him, it is "the early detection of those signals that could lead to strategic surprises and to an event that has the potential to jeopardize an organization's strategy".

On a more proactive note, weak signals mean that today's information can foretell the changes in the future. Nik Baerten, future explorer at Pantopicon, a studio of future explorers based in Belgium, expounds on the subject on their blog: "Change often starts with a ripple before it turns into a wave washing over us. Spotting signals of change early – when they are still murmur on the sideline – often means a strategic advantage, if one can interpret the signal correctly, anticipate and act upon it successfully".

// DETECTING WEAK SIGNALS

Weak signals are most often used in predicting new trends in technological and social development. As a part of her PhD thesis at Aalto University School of Economics, Elina Hiltunen conducted a global study about information sources of weak signals for future professionals and people interested in the future in general. "Interaction, openness and discussion were emphasized in finding weak signals. More generally, keeping eyes open, having sensitivity to changes, creativity, receptiveness, intuition and a curious mind is needed to find weak signals of change. Personally, as a weak signal hunter, I am very interested in changes about all aspects of life. Because I am an engineer, technology is especially fascinating to me. So, I spot weak signals everywhere using my camera phone". She then adds, in her article *The Future Sign and its Three Dimensions,* that the "Internet, especially blogs, remains among the most valuable hunting areas". In her interview with Nik Baerten, published on Pantopicon's blog, Hiltunen warns us that: "while you can get a lot of valuable information from the Internet, you have to compare or match that information to reality, to that which is really happening. Still, both the Internet and reality as we live it in our daily offline environment remain important sources for future changes [aka weak signals]". She continues by adding that people often ask, 'how do you know that a weak signal is relevant or not?'. "You don't know! If you knew then it wouldn't be a weak signal anymore. So, this is why I advise people not to filter things too much and to remain open to change!"
Francis J. Aguilar, Professor Emeritus of Business Administration at the Harvard Business School, defines environmental scanning as "an activity for acquiring information". He continues, in *Scanning the Business*, written in 1967,

"scanning involves simply an exposure to and perception of information. The activity could range from gathering data in the most deliberate fashions by an extensive market research program – to undirected conversation at the breakfast table or the chance observation of an irate housewife throwing your product into the trash barrel".

Moreover, for anticipating changes, it is important to look for emerging issues and weak signals of them all around us. This is called environmental scanning.

In order to find out *where* to scan for weak signals that can anticipate the future it's important to understand the logic of change. Sherry D. Ferguson, Professor at the Department of Communication of the Faculty of Arts at the University of Ottawa, in her article from the Canadian Journal of Communications on *Strategic Planning for Issue Management: The Communicator as Environmental Analysts,* argues that "few changes in the environment occur spontaneously: they start as ideas. These ideas eventually obtain a public expression in the press, radio, television, university conferences, and scientific journals".

A BIT OF THEORY

I felt like I needed some theoretical facts to get to the bottom of the weak signals concept. As I studied semiotics in great detail at University, I was excited to hear Hiltunen's theory on weak signals related to semiotics (aka the 'future sign'). **A weak signal is a signal of an event that revolves around a meaning.** The famous semiotician Charles S. Peirce used to say that nothing "is a sign unless it is interpreted as a sign". Hiltunen based her model of the 'future sign' on Peirce's concept of a three-dimensional sign – the sign, the sign's subject (aka object) and the sign's meaning. Applying Peirce's model of a sign to Hiltunen's future sign, we have three corresponding dimensions:

1. **Signal** – the observation of what is happening
2. **Issue** – the thing in itself, the concrete happening
3. **Interpretation** – the sense made out of the issue for the future

A signal is what we see and an issue is what is happening.
Elina gives a good example from the *Helsingin Sanomat* (Finland's main newspaper) that published news about H&M, the famous fashion retailer, who started to sell clothes under the vintage label (1. signal). Only 1% of their shops were doing that (2. issue). So from the point of view of the future sign, the amount of visibility of the signal was huge (which makes it tempting to name this a strong signal). But the reality (the issue: only 1% of H&M shops) was small. The meaning of this for the future was unclear at the time: will this become a bigger trend? In fact it has.
So, if we had assessed Hiltunen's future sign theory using Peirce's model in the H&M case, the signal was strong but the issue itself was rather weak. But stores such as American Apparel and Urban Outfitters have since successfully adapted (3.) the interpretation – the placement of selected vintage in their stores. Even if the issue seemed unimportant at the time, it gained momentum but this doesn't apply to every situation.
I'll give you another scenario. Do you remember in November 2009 "Google Wave: lets you communicate and collaborate in real time"? It promised to completely revolutionize our online communication. We were supposed to be able to mail, chat, share files and videos and work on the same platform. What I remember most is the crazy amount of time it took to get a user name for the tool. You had to wait for weeks just to get an account. I boasted to my friends when I finally got in and then, nothing! Nothing happened. Later on, in August 2010, Google announced that they "don't plan to continue Wave as a standalone product". It started with a bang but it quickly faded away. I will leave the reasons why it failed in the hands of the techs and keep my senses on for Google+ now. This experience taught me the valuable lesson that when it comes to weak signals, sometimes, you have to watch out for hype!

// WEAK SIGNALS – EMERGING ISSUES – TRENDS

There was a need for a little break during the interview and I decided to take a short walk. Helsinki is an amazing city and walking in the cold refreshed my brain. New questions arose! What makes a weak signal become a trend? If this is a pre-step to trends then is there a collection of boxes that you have to tick in order to be sure that you are finally dealing with a trend?

I found out that **weak signals are not going to become emerging issues because they are only signals of the emerging issues. Usually, the number of times that you identify an emerging issue in numerous contexts, thus forming masses of emerging issues, is what transforms them into a trend.** Miniscule, apparently insignificant stories can turn into significant cultural shifts and important societal trends. I'll give you an example that Hiltunen mentioned during our conversation. We all know that some people have stationary bicycles at home even if they are actually being used as clothes hangers. Today, in Second Life, your avatar can have a stationary bicycle too. In a small village in Finland, inside a McDonalds, children can pedal their own avatar in a computer game from a real stationary bike. In Vilnius, Lithuania, a new movie theater was opened where energy is generated for the projection by bike pedaling spectators.

All this information could seem unimportant individually but if you combine it and then ask yourself whether these new adaption of stationary bicycles could be something big in the future, new projects may pop-up. How do you think the guys came up with the Velo-Movie Theater in Lithuania? They kept their eyes open, watching what was going on around them, before combining what they had observed with their own ideas. So, as I understood it, the logical order of the birth of a trend should be: weak signals-emerging issues-trends. JewelEye from Netherlands Institute for Innovative Ocular Surgery is definitely a weak signal. To sum-up their information on the website: the desire to adorn our body exists since the beginning of time. Jewelry, make-up and more recently tattoos and piercings are accepted as forms of body cosmetics. But no decoration was available for the eyeball itself. So, in 2002 the Netherlands Institute for Innovative Ocular Surgery developed a cosmetic extra ocular implant (CEI) that can be implanted within the superficial, interpalpebral conjunctiva. The implant does not interfere with the ocular functions and is made of a specially designed material that can be molded in all kinds of desired shapes and sizes. This is not a trend – this is a weak signal. For this to be a trend you would have had to know somebody who has actually had the technology implanted already. Installing eye jewelry is something that not everybody can fathom. Most likely most of us won't be implanting a JewelEye in the near future but this information might lead to an 'emerging issue'. This innovation may not mean much right now but if you take the signal further, it connects with the broader idea of inserting technology into our bodies. This transforms the topic of technology inserted into our body from a weak signal to an emerging issue. Observing and spotting early adopters' reaction to the JewelEye is where things could get interesting because they have the power to transform the emerging issue into a trend.

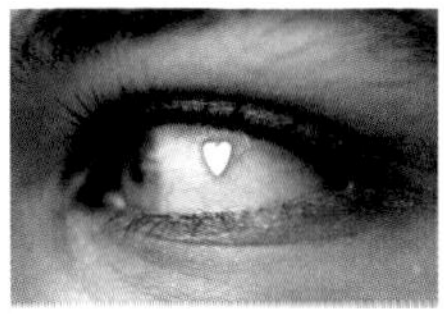

JewelEye
/Netherlands Institute for Innovative Ocular Surgery

The next big thing creating a buzz is the bionic contact lens. The bionic contact lens is already being developed by researchers at University of Washington in the USA to provide a virtual display that could have a variety of uses from assisting the visually impaired to the video game industry. "The device will have the form of a conventional contact lens with added bionics technology and the lens will eventually have functional electronic circuits and infrared lights to create a virtual display", says Wikipedia. In more simple words, this could mean replacing our PCs and laptops with a contact lens and accessing the real world through it. We could always be connected to the virtual world via an Internet connected lens. Imagine that!

Time is a big influence. A trend can come from a collection of weak signals but it needs time to develop. Weak signals are not hard to notice – as they are concrete events – but they are difficult to value.

Most of them are just overlooked in the business world because they represent uncertain knowledge. In the fear of risking failure, people are afraid to even mention weak signals. In some companies, it is irrelevant to talk about 'installing eye jewelry into the eyeball'. It is often considered more convenient to talk about climate change and globalization. "The truth is, however, that the big changes are already common knowledge. It is the innovations and events that bring fresh insights and, at best, create a competitive edge", says Hiltunen.

// TRENDWIKI

Making sense out of a massive amount of weak signals is facilitated nowadays by community tools or text mining applications. Elina Hiltunen told me about TrendWiki, a tool she developed together with Data Rangers, a Finnish company specialized in monitoring and analyzing information. "There is an apparent weak signals tool gap that TrendWiki seems to be helping to bridge". This tool is designed for organizations in order to help them collect and analyze weak signals. It utilizes the ideas of social media and includes sophisticated text mining tools. You might say, 'at last we have technology and I don't have to do this difficult task on my own!' But TrendWiki doesn't analyze, it just assists in the gathering process, making it easier for you to analyze. Every employee inside an organization should have the possibility to collect and share weak signals, thus giving them the capability to act as an antenna for future change.

Also, TrendWiki includes a toolbar that employees can embed in their own Internet browser. When they see an interesting webpage about some type of innovation or a new thing, they just click the button and add in a few details about the page. The signal is then saved inside the database, which can be browsed for specific kinds of signals by keywords, tags, dates or location of the spotted signal. The tool also features pages where employees can combine signals into possible trends. The software itself also does this. With its text mining feature, one can analyze texts for various types of things, such as connections between trends and keywords.

Your next question should be: how do you analyze the weak signals? It's actually best not to attempt to analyze a single weak signal, but to collect numerous weak signals. Then try to make sense out of them by uncovering patterns and clustering them. The key is to be broad-minded and use no filters in the early stage of the process. "One has to try to understand the noise. Only then true weak signals (of value) can be spotted", says Hiltunen.

// THE FUTURE OF WEAK SIGNALS

In 1998, Elina Hiltunen started working with this concept and people were asking what in the name of God it is. Today, in Finland, weak signals are highly valued as key tools to anticipate future changes in the business environment. Many Finnish companies have shown interest in the TrendWiki tool and some are currently using it. Even the countries' Ministry of Trade and Industry uses TrendWiki. Finland's politicians are already talking about weak signals. This is actually incredible since politics is a field that's usually late in terms of adopting innovative products and practices. Recently, a couple of books have been published about weak signals by Mika Mannermaa, President of Futures Studies Mannermaa and (in Finnish) by Sissi Silvan, CEO of Lollipop Management Consulting. I believe if we look at Finland as an example it's safe to say that more and more organizations will incorporate the use of weak signals in the future both there and abroad.
What I learned in Helsinki is that, when it comes to strategy work, weak signals are indispensible. Weak signals are the starting point of a trend; they are at the ground level, the first layer of a trend. Collecting and combining them into trends enable future scenarios. With these scenarios you can test your strategy and try to make it more flexible and agile for the future. You adapt. So, don't get boiled like a frog! Jump when you see the first bubbles!

This leads me to TrendONE's main office in Hamburg. I am curious to hear what people have to tell me about microtrends, the first level in the world of trends. Let's see...

If you are ready
to play Elina Hiltunen's
Futuropoly Game jump to page 159

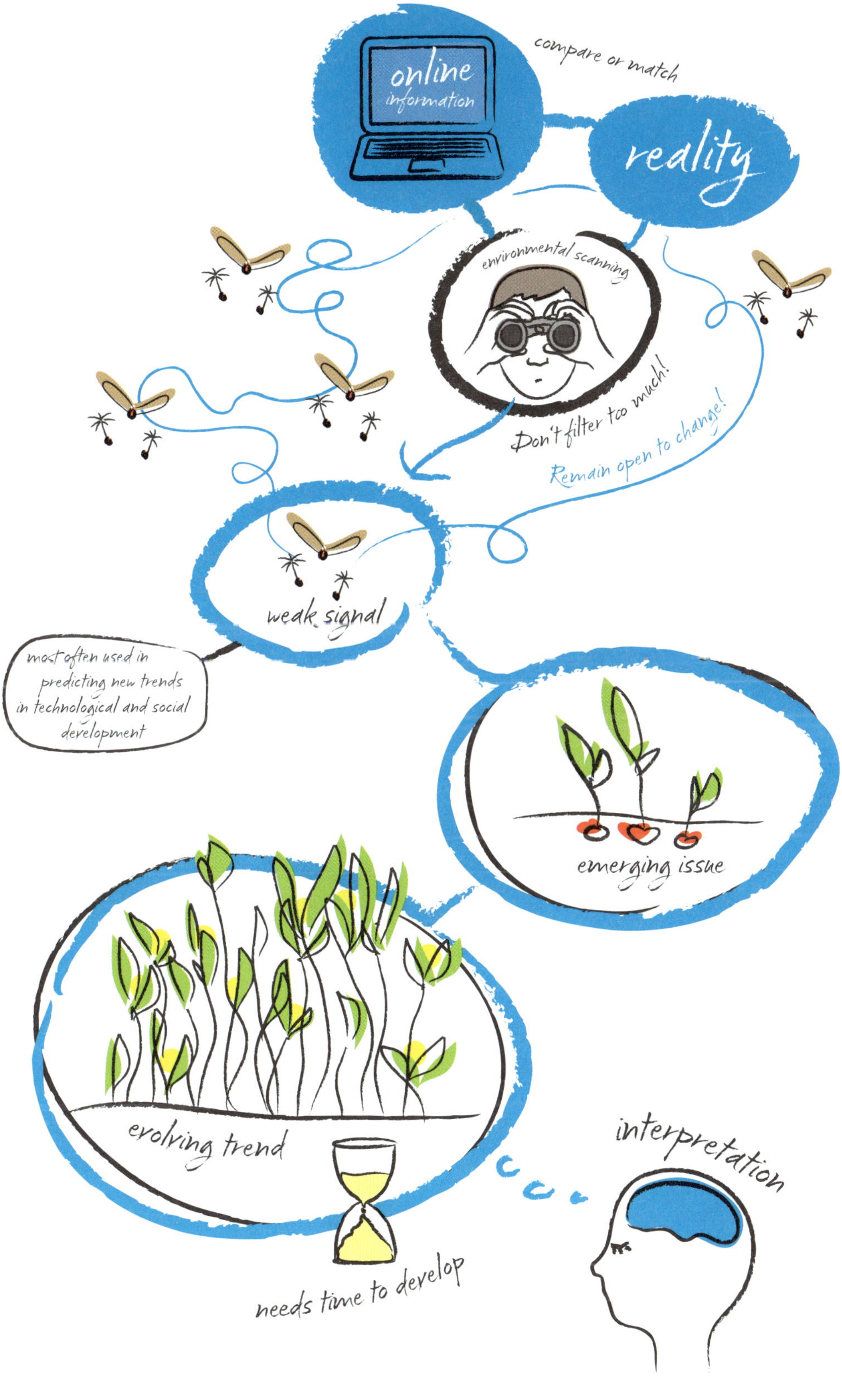

online information
compare or match
reality
environmental scanning
Don't filter too much!
Remain open to change!
weak signal
most often used in predicting new trends in technological and social development
emerging issue
evolving trend
needs time to develop
interpretation

MICROTRENDS

Nils Müller

Back in the Hamburg TrendONE office after my trip to Helsinki, I go deeper into the topic of microtrends with my mentor, CEO and Founder of TrendONE, Nils Müller. Meeting and working with him always turns out to be somewhat experimental and leads to creating outstanding future projects with high expectations and a great deal of vision. Most of the time, these projects can seem too daring for the weak of heart but always prove to be pioneering work. That's why I try to be around him as much as I can. Nils is a very busy person and I know I have to shoot my questions quickly. In the last minute he says: "Let's do this over lunch, ok?" So, I attack the topic as soon as we leave the office, heading for a fish dish in the harbor.

I know what weak signals are and the difference between them and trends. To recap the previous chapter in a nutshell, a collection of weak signals can form a microtrend. The difference between them, in terms of visibility, lies in their popularity and whether or not the specific idea was spotted in more than one context, industry sector or culture. But how do we define a microtrend? This is what I wanted to know.

Around 1999, Nike set up what we call today the 'customization' trend, under the name of NikeID. I was wondering how it all started. In the mid 1980s, Nike launched the Air Jordan, pioneering the ground-breaking collaboration of unique footwear for a famous professional athlete. We can call this a microtrend. "But the moment when we spot them", Nils points out that "microtrends are actually just examples of concrete innovations. They become trends, (or maybe not), with time. They are new born babies, at the micro level. For sure, they can develop and grow bigger with time".

But how?

More than a decade later, Glen Marshall, member of the Nike creative team, was out shopping with his son, who told his father that he was frustrated because stores never had the color shoes he wanted. That is when it all clicked. By 1999, NikeiD software, which allows anyone to personalize and design their own sport shoes was up and running. This chose your own color and/or print material customization allows for a seemingly endless amount of different designs to be created. The success of NikeiD made the

VIP customization microtrend available to the masses. It made it tip and become a macrotrend, a long lasting movement. Now with the release of the The iDNation, Nike has taken the concept a step further in the direction of community customization. Users can create their own sneakers in the 'StyleLab' and put them up for sale. Every creative shoe is shared with the Nike community, ranked and possibly bought by people around the world.
Today, other clothing brands offer the possibility to design your own products like VANS 'off the wall', miAdidas or Your Reebok. The product customization concept is present on the market in more than one industry and it is part of the megatrend called 'individualization'. Many of us have always wished to be different, special, unique. No matter if it is fashion, food, automotive or jewelry, customization is available for the mass market in order to make us feel individual. It all started with an impulse, an idea some time ago, a microtrend. The normal trendsetter reaction to a phenomenon such as this is to search for exclusivity. For these people, Nike has introduced 21 Mercer in Soho, NYC. Taken from their website: "Another unique customization experience available only at 21 Mercer is called Bespoke. Bespoke is available with the benefit of 1 on 1 consultation from dedicated Nike designers and offers an unprecedented opportunity to craft personalized, premium footwear-one product at a time".
If you pay attention to the signals around you on the market, you can get a pretty good picture of what is going to happen in the future. And if you are one of the main competitors in your field or if you want to be, this information is worth a great deal of money.

I found this information on swooshkidjm's flickr account. Check it out for the whole story behind NikeiD's birth.

// *TRENDONE'S APPROACH*

We arrived at the harbor restaurant and I wanted to know how somebody decides to start such a special business. TrendONE was founded in 2002 and at that time there were similar companies on the market but they were all focusing either on megatrends or on consumer insights. So, these companies were talking about cocooning, globalization and female shift, demographical change and so on. Nils realized this might be too abstract for a customer to directly implement in his daily business. "Let's say you approach BMW and you tell them about Wellness being one of the big trends of our time. They will probably say 'wow this is so interesting but what does this have to do with us?' This is why they need concrete examples to look at and to learn from", explains Nils. His idea for TrendONE was to be less abstract and more explicit and to focus on innovations happening all over the world. "We call our method 'microtrends focused research.' This was the initial idea for setting up an international trend scouting network that spots microtrends. And their findings are the best practice inspirations for our customers".

Today, TrendONE Germany is Europe's leading microtrends research company, daily scanning the relevant sources in 22 languages for innovations and microtrends from a variety of business fields. They look for intelligent marketing campaigns, new technology, media and online innovations, new business models and innovative products. Every month, the company publishes the most pioneering and trendsetting of the innovations they find in their concise Trendreports and Trendexplorer online trend database. Their objective is to provide businesses with a regular, comprehensive review of the latest developments on the market and to offer a boost of inspiration.

As Nils says, "the future will only come to those who believe in it!" Meaning that the companies who follow the market developments, who are open to thinking out of the box and who recognize today's key trends, will quickly adapt to the future and reap its rewards. TrendONE continuously inspires advertising and creative agencies, as well as internationally renowned brand names with the information they offer.

WHAT IS A MICROTREND?

I was now wondering: when it comes to microtrends, what are the decision makers? What are their characteristics and what makes somebody admit the specific information is a microtrend? Briefly, "for something to be a microtrend it has to be new, intelligent, a business model, mass market-ready so you can scale it and it must change the structure with which we do business, communicate, network and live our lives", says Nils.

Let me give you an example of a microtrend coming from TrendONE's innovation database: at the end of 2009, the American software developer Swartz Enterprises launched an iPhone app which activates the phone's camera as soon as it identifies a person's face. The app uses a special algorithm that can recognize facial characteristics even in difficult conditions, such as poorly lit rooms. Then think about this *old* technology applied today. Just yesterday I read on psfk.com about the JR EAST railway company of Japan that has set up a vending machine in one of its central Tokyo stations that can detect a customer's sex and age and recommend drinks based on these factors. The system has a 47-inch touch panel and a camera for face detection; once the customer stands in front of it, it displays the recommended drinks highlighted with a red star. **So, suddenly, from an old idea a new product came up. This should explain what innovation contagiousness means, right?**

TrendONE believes a microtrend is an innovation that was not on the market before. It is not an evolution of a specific product, like a bigger display or a faster processor would be, but it is a real shift.

Another criterion for something to be a microtrend refers to being a business model and therefore scalable. TrendONE only delivers innovations that are happening now and can be applied to your business starting with tomorrow. As the information is a concrete innovation, customers can directly and immediately work with it as they also find out from the monthly report what company produced the specific innovation and who to contact for starting to collaborate with them on projects. Super efficient.

In conclusion, the collections of innovations that are published in TrendONE's database will always be the latest things that just happened somewhere in the world and each of the examples are unique. This means that the innovation database is not a collection of all the campaigns, not an overview of what's happening on the markets but it is filled with things that nobody has ever done before. Therefore, you will not find the same idea twice. If, let's say Diesel introduces the Facebook web cam in their stores and then Levi's is doing the same, the microtrend will only be posted once, in reference to Diesel because it was the first to do it.

EYES WIDE OPEN

Remember my first meeting with Nils at the Ritz? After that, while I was researching and interviewing companies for this book, I was trained to be a trend scout at TrendONE. I studied coolhunting and worked as a trend scout independently, so, I was familiar with the topic. There are smattering of trend research companies in the world working with collaborators and for the most part in a similar way, but I'm going to narrate my version of the story.

TrendONE's main investigative tool is the scouting network. Trendspotters work for TrendONE as freelancers. They look into their local markets for what TrendONE refers to as microtrends. The company offers them an online trend scouting tool where they can easily drop their microtrends suggestions. The trend analysts check the innovations during a trend workshop, on the criteria Nils mentioned above.

microtrend

Just in case you forgot:

1. new
2. intelligent
3. mass market-ready
4. structure changing

Microtrend research uses almost the same methods and tools as weak signal research but the objective is different. We can deduce this from the descriptions of both concepts. All the innovation management and future research companies take their information predominately from available, free sources. There is no secret. Everybody has access to this information; they just need to know how and where to look for it. Then what makes it so special? Why can't anybody just do it online by himself? Well, it may not be rocket science but such a service delivers qualitative information in an extremely short time. No newly made, understaffed, inside company department could quickly reach this point. The average company couldn't afford the time and money for all the employees needed to research, analyze and produce trendreports on an on-going basis. The team at TrendONE practically breathes innovations every minute of the day, and therefore, they have a trained eye for what is relevant and what is not, in today's context. Moreover, one person could not do the work of the 80 trend scouts that are spread around the world, with varied backgrounds, experiences and local insights. "We do innovation scouting. So we are the ones who know first when there is a new technology on the market somewhere in the world. It's our vision to be the fastest and to offer personalized services to our clients", says Nils.

// WHAT CAN YOU DO WITH MICROTRENDS?

Ok, but then what do you do with all these microtrends? How can you extract innovation out of them? There are a variety of ways, business wise, to use microtrends. TrendONE believes in their relevancy for the industry as an orientation instrument.

THE RADICAL GAME

I have attended some training sessions held by pretty famous companies, all promising to make you a better manager, a more organized person or a more creative one. I cannot say that I ever left with a new idea from these sessions. The Radical Game is not just another workshop. What is new here is the technique. The objective of the Radical Game is to develop new business models and products in an open, creative game situation. At the beginning of each Radical Game a warm-up session takes place in the form of a presentation of the most inspiring Radicals (aka microtrends). The idea is pretty simple. All the microtrends are printed out as cards, as it would be a real card game. The participants are then split into groups, mixed, with people from the company, from different departments and from TrendONE. They all start playing with the innovations, combining them. For example, a Nokia radical game could set as a goal creating the next generation of smart interfaces. One of the players can come up with the visual control technology and another with the voice processing technology and a third one with mobile holographic technology. Using a combination of these three innovations, you could come up with a new idea, a new innovation and you own it! **Innovation goes through the existence of innovation. It is all about playing with it.**
Finally, the players write down and draw their ideas on a piece of paper and present their best one in 60 seconds. The evaluation is the fun part. Every player receives virtual money and has to 'put the money where the trend is' – bet on the idea they liked the most, the one that they would like to implement.
Sure, this is just one out of the tens of creativity techniques out there.

The difference here, as far as I am concerned, is that you can have an idea very fast, inspired by an already existing brilliant idea, based on existing technologies that you can already use. Moreover, at the end of the workshop, after the evaluation, you clearly know which of the innovative new ideas could be part of your next project.

As Nick Sohnemann, Innovation Advisor at TrendONE, always mentions, "people don't sit on their toilets at home waiting for inspiration to come. People work hard and read a lot to get inspired". Everybody should understand that, as we are all different human beings, getting our inspiration from the same source doesn't return the same idea. An idea is always tailored to your mind, perception, brand, knowledge and so on. This is why a poly-social group of people brainstorming on the same topic will bring infinitely more innovative ideas than a group that belongs to the same industry, works in the same department or has studied the same subject.

TrendONE also conducts an analyzing process when the team clusters microtrends and detects new, longer-term trends, so-called macrotrends. Building macrotrends means grouping microtrends, analyzing their deep significance, finding their origin and bringing them together into a new concept according to whatever it is they have in common and is relevant for tomorrow's society. For example, the next generation of technology is about gesture control, voice control, holographic screens etc. Products are not high-tech anymore, with lots of buttons and looking rough, but they are 'shy', have simple interfaces and make technology seem like it hides away. TrendONE calls this 'ShyTech'. The iPhone or the Nintendo Wii have very complex technologies inside but as a user you never interact with it, you just get in touch with the simple, very easy to use interface. One topic from last year, also published as a white paper that can be downloaded for free online, is the 'Outernet': "Under the name 'Outernet' a technological development is approaching which will profoundly change our relationships with each other and with the objects around us in the world. The Internet is leaving the previously detached realm of cyberspace and placing itself over our environment like a second skin".

KEYNOTES are also an innovation tool. At TrendONE, keynotes are 40 minute presentations where you are taken on a multimedia, video-based journey into the future, live testing the innovations and microtrends. Cases presented via live demonstrations, samples and videos make TrendONE's keynote presentations very different from others in this field. You are no longer sitting in front of static, boring power point slides with bullet points. Some of the current topics for the workshops maybe '2021 Time Travel', 'The Outernet – The internet is exploding into the real world: Web of Things, Augmented Reality, Hyperlocality', 'ShyTech – The success factor for the future of technology'. Practically all of these are macrotrends and megatrends, clustered from the microtrends scouted monthly are presented into the reports and database.

// EVALUATION

We have slowly started walking back to the office after our delectable lunch, but the matter of 'evaluation' suddenly appears in my mind. I started tossing questions at Nils. He takes a seat on a huge grey stone by the side of the road, and tries to field them: "TrendONE's value is inspiration and coming up with new ideas by delivering an expansive and insightful market innovations overview. You need this to open your mind, to be inspired and to think of complementary tools for your own business. You need someone who tells you all these things". When it comes to creativity and idea ownership, people and brands are not sharing enough. As most of TrendONE's clients don't say how they developed a specific idea, it is difficult to find out whether it was TrendONE's success story or not. For example, TrendONE found the video in print technology very early and reported this to their clients. A couple of months later some of their clients had it in their magazines and advertising campaigns. OTTO, the largest mail order fashion retailer, has integrated Augmented Reality features into the winter edition of its 'mytrend' fashion catalogue just after Johannes Dahlhaus, the Manager of Catalogue Advertising Unit 1 at OTTO, attended a visionary keynote presentation given by Nils. By holding the current print edition of the catalogue in front of a webcam, readers enter into the mail order company's virtual product world, where they can interact with the models on display and admire the products from a variety of 3D perspectives. This turns online shopping on your home computer into a wonderful new experience. "With the Augmented Reality application in the new 'mytrend' for winter, OTTO is providing an impressive example of how to enhance print products with added value for customers. This project has enabled us to collect some highly valuable experience that we will be able to use in future design concepts", said Dahlhaus.

// THE FLAP OF THE BUTTERFLIES

There is a story that I once heard during a PechaKucha session in Barcelona, a simple presentation format where you show 20 images, each for 20 seconds with your new idea. I thought to myself "this here is an example I can completely relate to microtrends and their effect".

In the early 1960s, a professor at MIT, Edward Lorenz realized that small differences in a dynamic system such as the atmosphere could trigger vast and often unsuspected results. Therefore, 'the flap of the butterflies' can cause severe storms on another continent. Likewise, microtrends cause severe market changes! Mark J. Penn further states in his book called *Microtrends. The Small Forces Behind Tomorrow's Big Changes* that "the one-size fits all approach to the world is dead". Penn, the well known pollster who identified soccer moms as a critical voting constituency in Bill Clinton's 1996 presidential campaign, identifies a microtrend as an "intense identified group, that is growing, which has needs and wants unmet by the current crop of companies, marketers, policymakers, and others who would influence society's behavior". The difference between the approach of the book you now hold in your hands and Penn's approach to microtrends stands in the fact that we are looking at the phenomena from different sides of the street. There is the market point of view vs. the consumers' point of view. Therefore, in my opinion, what the author calls 'microtrends' fits more with my understanding of consumer insights, which forms part of another chapter of this book.

In order to see changes, Penn says, "we need better tools than just the naked eye and an eloquent tongue. We need the equivalent of magnifying glasses and microscopes, which in sociological terms are polls, survey and statistics. He adds later, "I have used the most reliable device I know of to spot trends, or the shifts and evolutions in these groups. numbers".

My understanding of microtrends doesn't refer directly to the consumer or to numbers. It corresponds to the market itself and what is going on in terms of innovations, examples of behaviors, activities, new but marketed and accepted technologies (emergent technologies that are still in the form of concepts or prototypes would be weak signals) that have never happened before and therefore cannot form an "intense identified group".

A microtrend is indeed something growing but it is not yet completely visible or influential. It is still on the curve, climbing up, growing to reach the tipping point (or not), hit the markets and become a macro-trend that is influential to society. If and when it will reach its goal we do not know.

Microtrend research is a qualitative, empirical, no-numbers-involved method as opposed to consumer insights research which sometimes takes into consideration the quantitative side as well. I learned that innovations and trends these days are boosted by product and technological developments and not so much by the consumer anymore. **I am wondering, what are the limits of trend research?** Well, the uninteresting thing about trends is that everyone follows them. Really cutting-edge businesses like Apple don't follow trends at all. They are aware of trends. They invent things that they think are good ideas and sell them, find markets for them. That's an extremely important point. The world is moving at a very rapid pace; brands have to be one step ahead the consumers' needs if they want to

innovate. Meeting customers' needs should no longer be the ultimate goal, as today's client needs to be always surprised with something new, something he never thought of. A combination of sources and approaches will probably generate the best results.

// THE MAGIC OF 1% OF THE POPULATION

I like "the magic of the 1%" of the population concept from Penn's book on microtrends: "In today's mass societies, it takes only 1% of people making a dedicated choice – contrary to the mainstream's choice – to create a movement that can change the world". As I said before, Penn's method is to use a combination of survey and opinion polling data. For instance, the average US commuter is estimated to have a daily 25-minute journey to work. This compares to the European Union average of 38 minutes and UK employees who travel an average of 45 minutes. However, according to the US Census Bureau, in 2000 there were 3.4 million Americans, whom Penn dubs the 'Extreme Commuters', who traveled daily for at least 90 minutes each way. This population had doubled in a decade. The conclusion is very nicely stated: "By focusing only on the major trends that reach a tipping point, most observers are missing the fact that you no longer have to reach that point to be a successful trend with a vast potential impact on society".

Over 3 million people – "the magic 1%" for a microtrend – are waking up at the crack of dawn, crossing state lines and even weather zones to get to work. This example (taken from *Just 1%, The Power of Microtrends* by Mark Penn &E. Kinney Zalesne, from Change This No. 38, an online community website for spreading ideas) proves how risky it is to make assumptions. "While you are focused on big trends, microtrends can seep in and upset your expectations", says Penn. He adds that "the explosion of choice has,

in turn, created hundreds of small, intense communities defining themselves in new ways. So now, we can no longer understand the world in terms of a few mega forces sweeping us all along. Rather, society is being pushed and pulled by microtrends – small, under-the-radar forces that can involve as little as 1% of the population. Microtrends are about the niching of society".
I couldn't agree more with Pero Mićić, founder of FutureManagementGroup, and his view on trends from his wonderful book *The Future Radar*. He says that, looking from a strictly semantics perspective, "the popular understanding of trends is not really about trends at all. The observation of innovations described in this way is, at most, signals that could, but do not necessarily, refer to a real trend. When looked at more closely, both the popular and the phenomenal understanding of trends concentrate much more on the present than on the future".
Does this confuse you even more? What he means is that **trends are observable developments in a relatively clear direction**. This makes things easier, more concrete. It links with what I talked about in the previous chapter with Elina Hiltunen: the transformation of weak signals into microtrends. You would have to know somebody who is actually living the phenomena or using the product or the service to call it a trend. Therefore, a microtrend doesn't refer to the future but to what is happening NOW.
Why is this helpful for harvesting innovation? Well, because microtrends are signals of possible future developments. Ignoring them, as a company is like giving yourself the death sentence.
In 1996, Andy S. Grove, Chairman of the Board of Intel, wrote a book called *Only the Paranoid Survive* where he uses the term '10Xforce' (meaning, ten times more powerful force) to define the time when a business has to transform under a super competitive force. Like, for example, how the Internet transformed businesses. Chell Smith, CEO of Capgemini, argues in the article *Microtrends – Today's Business Paradigm* published online on GDS International's website: "I believe business executives today are forced to navigate through even rougher seas. Instead of an infrequent 10Xforce, they must now deal with more frequent, smaller and faster shocks. I call

them microtrends, smaller trends – specific to an industry, a technology, regulation or other disruptive force – that continuously transform businesses". She continues to say in the same article that microtrends "pose challenge after challenge" and that anticipating the next microtrend is or should be a primary function of the top management from any company. Managing its impact in timing and size adds to the task.

In today's business context, CEOs no longer have time to do things the traditional way, building a business plan that is updated quarterly: in a microtrend environment this process falls apart.

Magnus Lindkvist pointed out to me something he once heard: "when I see an image of an atom bomb I feel nothing but when I read the thoughts of a single fisherman he can move me to tears". This is the secret. Big movements are abstract and people can't relate to them.

// THE FUTURE OF MICROTRENDS

I asked for Nils' thoughts on the future of the 'microtrends approach' to building innovation. So we stop in the stairway to discuss this topic for a bit before heading back to the office. He starts telling me that the futures industry is at its early beginnings. There have been futurists before, looking into a crystal ball 'guesstimating' what is to come, but now future thinking is craving standardization and tools are being developed that can make it happen. In the past few years we have witnessed companies inventing innovation departments, coming up with future research methods and employing trend spotters. Innovation means also marketing because traditional direct market research has no use anymore. What can we do to surprise the customer, what can we develop that is new for them? Of course, the difficulty lies not only in the scanning process but also in building a good market strategy for it. Questions related to 'good time to market' (whether the market is ready for adopting an innovative product and incorporating it into society), what are the market's adoption limits etc., are part of strategic decision-making. The 'location based' strategy concept has been discussed for over 6 years and only now, the market is capable of receiving it. Failing is not so hard and finding the right time is damn difficult. This is why scouting and microtrends are just a small part of the big futures process. In Mark J. Penn's book, he does not predict that all 75 microtrends presented there hold equal weight or that each of them will have the same type of impact on our future. Instead, microtrends should be understood more like a radar.

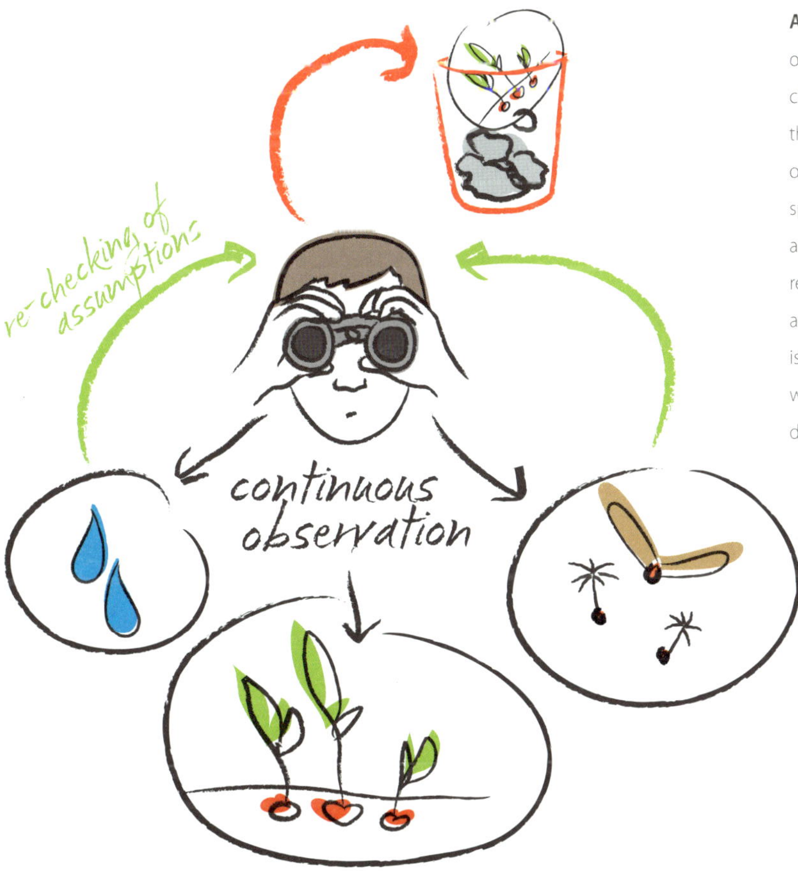

A good observer of trends is characterised by the continuous observation of specific issues and the constant re-checking of his assumptions, and is always someone who is good at discarding things.

Peter Bisson, CEO of McKinsey & Company – a leading management consulting firm advising companies on issues of strategy, organization, technology, and operations – once said that "the reason why it's important to look at these forces now, and really at all times, is because they happen so slowly that it's easy for them to slip to the bottom of the inbox.
Each day, each quarter, there will be a more pressing, immediate concern. But you wake up in five years' time or ten years' time suddenly finding yourself facing competition that is very well entrenched. And so, on any given day, this didn't make the top of the list, but it was tenth on the list every day for a decade. By then it becomes, perhaps, a force that's unstoppable".

After spending this amazing afternoon with Nils I feel like I know enough to travel to the next station of this road trip. Then he asks me, "Do you know trendwatching.com in London?" I say, "Yes, have been following their monthly brief for quite a while" and then he quickly adds "Ok, that's good but not enough. You should meet them. What are you doing next Monday? Let's book your flight to London".

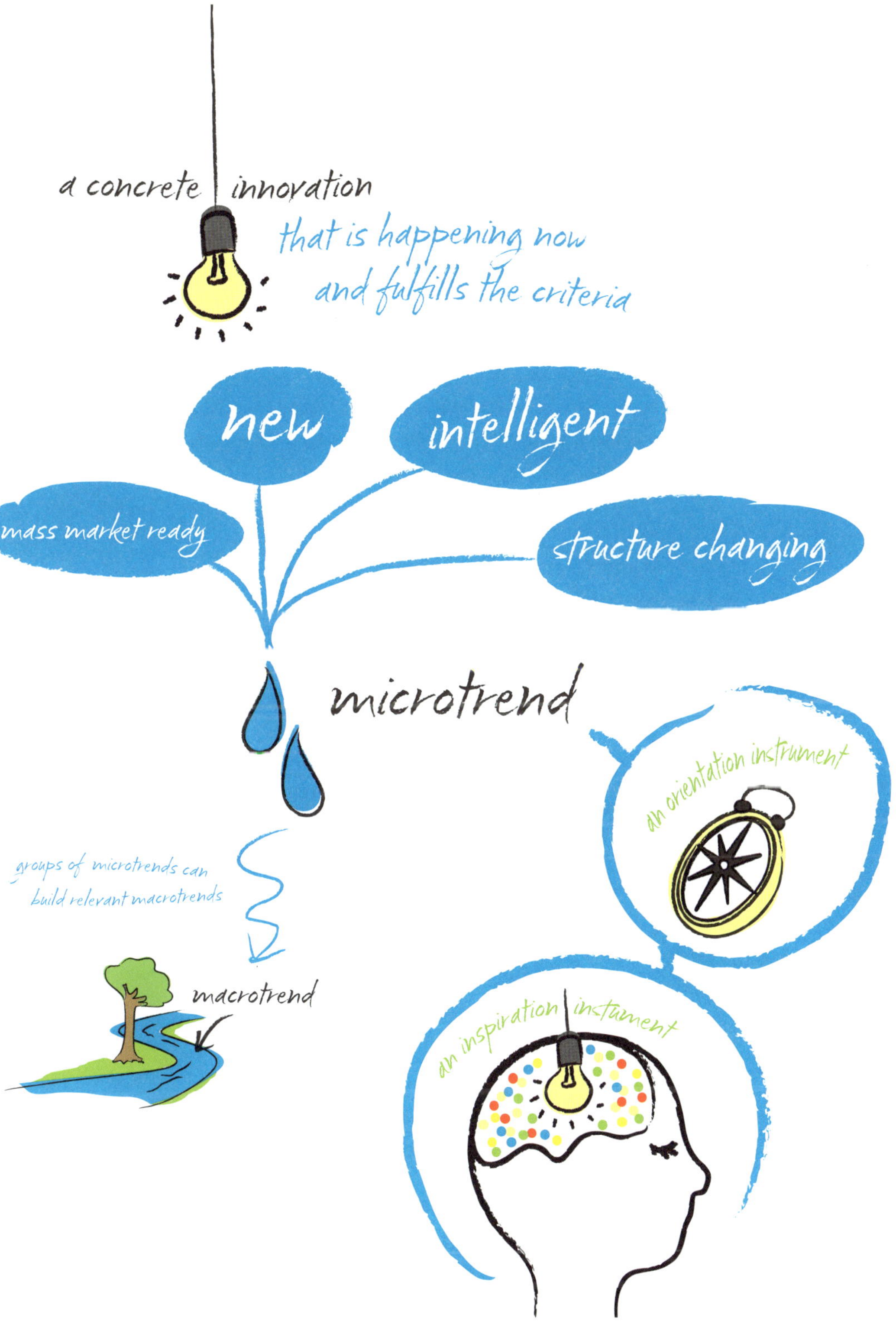

a concrete innovation
that is happening now
and fulfills the criteria
new
intelligent
mass market ready
structure changing
microtrend
groups of microtrends can
build relevant macrotrends
macrotrend
an orientation instrument
an inspiration instument

MACROTRENDS

To grasp the special characteristics of each trend layer I knew I still had a long way to go. I had my flight booked to London. The night before I took off, I had a classic case of 'monkey-brain'. That's a term my friend uses to describe those moments when you can't sleep because you're obsessed with something and it won't stop jumping around in your head. My obsession was reviewing all I happened to know about macrotrends, so I made a list:

> Macrotrends run deep and wide
> They embody major cultural shifts in how we live and what we want
> Weak signals and microtrends might come and go, but macrotrends endure: they are longstanding cultural shifts.

LOHAS is a macrotrend describing a purchase group that emerged gradually a few years ago and was interested in a sustainable lifestyle and ecologically correct living. The most important characteristics of this group are its available purchase power and its efforts with regard to minimizing consumption.

Some macrotrends are new, like Lifestyle of Health and Sustainability (LOHAS), while "others like Friends and Family, represent the very essence of human experience", says Iconoculture – a consumer research company.
Let me give you a microtrend example as part of the LOHAS macrotrend so you can better understand what they are about: the vegetarian restaurant chain Otarian just opened its first branch in New York City. The restaurant's concept places particular value on sustainability and environmental responsibility and therefore, the carbon footprint of each dish is clearly visible on the menu. The idea is to inform customers about what effect their food has on the environment and to encourage them to eat less environmentally-damaging meat and more locally-grown vegetarian foods. No products are used in the restaurant that have been air-transported and all the furniture is made from recycled materials.
Planning teams in advertising agencies conduct, among other types of research, environmental and trend analyses and they label the gathered information as macro (outside the brand and its immediate market) trends (likely behaviors). Jonathan MacDonald, Co-Founder of This Fluid World and Managing Director of JMA – a specialist execution agency in the new world of communications & technology – in his article, *The first 6 MacroTrends*

of the 21st Century from October 2009, simply calls macrotrends "the major changes happening in business". **A macrotrend has a long life and is a result of many forces coming together from our community that we cannot manipulate.** Like a river, you go with the flow or exhaust yourself fighting it.

A macrotrend can also be a business driver like a trend in the community that builds momentum and triggers an opportunity for your business. For example, a current driver for web meetings can be high travel costs. This can increase the demand for professional conference call systems. Global macrotrends have a strong influence on economies, people, the environment and society. Identifying and rolling with these trends is important to the future of companies, governments, families and investors as these trends are not your hot idea of the month, but rather changes that span years.

// @TRENDWATCHING.COM

How are macrotrends delivered to brand strategy teams? How is this different from starting your research for a strategy by spotting microtrends? I was about to learn all this from Chris Turner, Head of Research and Analysis at trendwatching.com, a consumer trends and insights company based in the Holborn area of London.

Reinier Evers is the founder of trendwatching.com. He was looking for new ideas and trends that he could take advantage of and exploit for building a new company when he suddenly realized he was spotting so many ideas and so many trends that this was an opportunity in itself: to provide a source of information on trends rather than exploiting one of them. At the beginning, trendwatching.com was offering a lot of free content, which they still do in their monthly briefs, but now they also offer a subscription based premium service. The briefing feels like a little teaser but even so, I believe there are many entrepreneurs who could gain insights just by reading the monthly issue.

Reinier Evers

Trendwatching.com today is very successful. Their business focus is on consumer behaviors (further clustered into emerging macrotrends), why they like what they like and why they don't like what they don't like and searching for cool new services, businesses and products.

Chris Turner

To me, this approach sounded familiar. To be honest, all trend firms say that they have a slightly different approach; this is actually true. I'll explain why. It all comes down to the techniques (**bottom-up – spotting micro to identify mega or top down – spotting mega to break it down identifying macro & micro**) and the shape that the information takes when it is delivered to the client: a microtrends report, macrotrends report or consultancy of how to apply these trends. "Every trend agency has its own competencies and I don't think clients pick between them but rather get as much information as they can from all trend agencies because the information is always slightly different", says Turner. My! That's a very interesting insight!
Established in 2002, trendwatching.com coordinates a worldwide network of analysts and spotters from its London offices. Its sister-site is Springwise New Business Ideas. According to them they are "independent and opinionated, one of the world's leading consumer trends firms, relying on a global network of hundreds of spotters". Their trends, examples and insights are delivered in more than 180 countries to 160,000 business professionals, help marketers, CEOs, researchers, and anyone else interested in the future of business and consumerism.
"Trends still can mean everything from 'Ageing populations in Central-Europe' to 'Spring 2012's skirt lengths.' Major consumer trends [aka macrotrends] are more like currents than one-time killer waves", says Evans. trendwatching.com is tracking consumer trends that are already one step ahead of innovations research (which is at a micro level) so we will refer to them as macrotrends in this book. Categorizing is the hardest thing to do. It's tricky. Many of the words from the trends industry are used carelessly by many people. At the same time various small communities have different ways of using terminologies, so the same word can mean different things. In fact, I am having great difficulties while writing this book, as there seems to

be not just a variety of definitions floating around regarding micro, macro and megatrends, but there's no real consensus on their spelling either. Trendwatching.com's monthly briefing is a combination of micro and macrotrends. "Rather than just being a micro/macro theoretical trend, it is about consumer behavior and a way in which that specific consumer behavior is manifesting itself", says trendwatching.com. For example, 'Statusphere', as an emerging macrotrend, was the subject of one of the last year's monthly briefings. Here are some excerpts:

> "Status has always been a key if not 'the key' behind consumers purchasing. The way in which consumers are now appealing to a status is more and more diverse. Status, as a buying driver, has always been there, like a forever trend. But today is different; it is evolving into something else. It isn't such a one dimensional phenomena anymore. It is no longer about what the brand can do for you but about what you can do with the brand.
> As consumers are starting to recognize and respect fellow consumers who stray off the beaten consuming-more-than-thou-path, 'new' status can be about acquired skills, about eco-credentials, about generosity, about connectivity... All of this makes for a far more diversified 'STATUSPHERE' than most brands and organizations have traditionally catered to. Time to really figure out how and where your customers are now finding their status fix."

METHODS, METHODS, METHODS

trendwatching.com is not a traditional research firm; they don't conduct focus groups and surveys. Instead, they explore the real world looking for opportunities for innovation. It sounds almost idyllic, but how does all this happen? Similar to TrendONE, they also have a core team of trend scouts. They appear in the form of an online community called Happy Spotters, available to whoever wants to join. So, every happy spotter periodically receives a brief from trendwatching.com. Here is one from last year:

> **Request #9**
>
> Maturity Audiences in mature consumer societies no longer tolerate being treated like yesteryear's uninformed, easily shocked, inexperienced, middle-of-the-road consumer. As a result, we're looking for new examples of brands using mature or 'edgy' products or services to serve these consumers. We would like to hear about truly original innovations that are more daring, more opinionated, or more risqué than ever before. Of course, we're not simply looking for truly shocking innovations, but instead for brands using mature content in a smart and transparent way – without alienating consumers. For an idea of the kind of examples we're looking for, take a look at Philips line of Sensual Massagers, which shows a large mainstream brand unafraid to develop and market mature products. Also check out Air New Zealand's 'Nothing to Hide' campaign, an example of a well established brand attracting attention with a daring campaign.
>
> **The usual guidelines:**
>
> - Any B2C industry will do, from any country in the world.
> - We're not looking for older examples: they really need to be very new!
> - Please send us specific urls that really refer to the actual product or brand you've spotted.

I guess now you've got it. From the brief you can also read trendwatching.com's criteria for a trend. Being a bottom-up approach, they develop and deliver macrotrends by working on the microtrends they receive from the Happy Spotters network. That's where they get loads of brilliant examples from. Then, the analyzing process begins..

The process goes something like this: the team sits down and talks, trying to spot trends emerging from all the information they receive, identifying something that is worth looking at more in depth. How? "It's an art really. We have our trend framework which is built into our premium service. What we aim to do is try to fit in every example in the framework. Therefore, the most interesting part for us comes up when something doesn't fit in the framework that we developed. This could mean there is something emerging, either a new macrotrend, or a new development for an existing macrotrend. So, the examples sort of lead the way", says Turner.

Talking to him about the criteria of a macrotrend, I started to become suspicious as it was not the first time I've heard that the information has to be "exciting, innovative, challenging, something truly different".

I couldn't help but wonder, are there different levels or scales of innovation? Yes, there are. trendwatching.com judges every example on its own merits. For example, the macrotrend entitled 'Branded Brands' is all about well-known brands coming together to produce something collaboratively. Obviously, the scope to be truly innovative with this trend is perhaps more limited because it's always about two brands (that we are already familiar with) coming together, but it is just a case of finding really interesting brands that are working together in unexpected ways. Sure there are areas in which you can be much more surprising than that, right?

All this talk about micro, macro, more detailed and less detailed made me think about industries. **If we have layers for trends do those correspond to industries?** Does fashion fit more with microtrends because it needs to change more quickly?

Do long-term brand strategies have to pay more attention to macrotrends because they have a longer time span and are more descriptive and safe?

Every industry has different priorities and needs to be aware of microtrends because they will always affect them. When it comes to macro or megatrends some are more relevant to some businesses than to others. "Consumer trends are bigger than microtrends because they are universal in a sense that anybody can tap into a consumer trend", says Turner.

I was pondering the reason why companies like trendwatching.com exist. But before I had managed to shape the question silently in my head, Turner answered it: "Macrotrends play a key role in shaping the consumer environment and we like to think that there are trends that kind of slot behind them and perhaps reflect the way consumers are reacting at the border". That's an interesting place to look for insights! Moreover, I realized that these kinds of companies win because they are much faster in delivering real information. Making polls and in depth research takes a lot longer than drawing conclusions from observable consumer behavior and technological developments. Trend research agencies speed up the process.

However, when it comes to the so-called macro contexts like economics, demographics, government, there are other levels to this game.

You remember what I said about Mark J. Penn earlier? He is combining consumer insights with polls and surveys. Similar to his approach, there are academic bodies, like McKinsey & Company who look in great depth at macrotrends. Most strategists feel that four areas need to be studied in detail: politics, economics, society (demographics) and technology before taking any decision regarding a change.

With this in mind a cheeky question was on the tip of my tongue. Should I, or shouldn't I? Ok, here we go. I asked why trendwatching.com's service is then valuable for clients. In a totally relaxed way, Turner explained: "I think it all goes back to examples, to our focus on the real world, to our focus on stuff which is happening now, that we've seen and identified. This brings trends to life and it provides a stimulus for brands to innovate. Reading the trend theory makes sense in an intellectual way, it satisfies the need to know what's going on. But in terms of innovation it is really the examples that help brands to innovate".

Then I realized, the difference between these two types of companies (trend research delivery vs. strategic consultancy based on trend research) exists not only in the final delivered service but also in the manner in which the client is assisted, all the way to the end (creating a new product, service, strategy) or just delivering the information (letting him choose what to do with it). So, there is no good or bad, no high level low level or questions of quality between these companies. It is just a matter of establishing the level where the service is delivered. The client simply has to choose what kind of service he wants to receive.

// THE FUTURE OF MACROTRENDS

Coming from the clients, there is a real hunger for more knowledge and insights. The requests for more specific services point to the development of the trend industry. Turner adds that "through the recession, as a trend firm, it has been interesting to see that we have an appeal that goes beyond the climate. During the recession, people are keen to know about trends because they can gain an edge in a difficult environment. But likewise, in times of good, people want to know about trends because they want to make the most of the good times, to tap into something which is part of the dynamic of the culture".

In conclusion, according to trendwatching.com, "brands should forget about 'Nice to Know' or 'Pie in the Sky' and start applying trends straightaway". They give four pieces of advice on how to make some money from the innovations they bring to light.

Turner ends our interview with an optimistic thought: "Trend research is an industry that is going to appeal regardless of the times, culture, location, or industry. It will have a lot of success".

"One needs to ask himself if they have the potential to (and if so, how):

> Influence or shape your company's vision.
> Inspire you to come up with a new business concept, an entirely new venture, a new brand.
> Add a new product, service or experience for a certain customer segment.
> Speak the language of those consumers already 'living' a trend".

As every trend firm is different, there always seems to be room for one more. Trend investigation and analysis will continue to grow in popularity, as it is an ever-changing, advancing field full of surprises. Later I find out that there is a growing desire to discover ways to tap into the Central and Eastern Europe's huge consumer markets. For me, it's interesting and somehow flattering to come from one of those countries in this situation. What I know is finally becoming 'gold' for the West.

many forces coming together from our community,
that we cannot manipulate
it's easier to go with the flow
macrotrend
longstanding cultural shift
significant influence on
economies
people
the environment
society

MEGATRENDS

There isn't anything fundamentally new about megatrends in terms of content because by definition they represent global behaviors and long term trends. Some have shaped our world for hundreds of years. Most have been triggered by a major change, such as a technological breakthrough or a shift in demographics. Ok, so if there's nothing new about them, what about the 'looking into the future' mindset that we want to follow throughout this book?

"Turns out that the analysis of these past developments shows how future megatrends are likely to develop", says Lars Kalbreier, Head of Equity and Alternatives Research Credit Suisse Global in the Global Investor Magazine 2.09. And that's where it starts to get interesting.

Actually, according to a CSIRO report (The Commonwealth Scientific and Industrial Research Organization – Australia's national science agency), *Our Future World: An Analysis of Global Trends, Shocks and Scenarios,* "Megatrends are collections of interlinked trends that will change the way people live and the technological products that they demand". Gitte Larsen, futures researcher at the Copenhagen Institute for Futures Studies (CIFS), writes in the article *Why Megatrends Matter* from Futureorientation 5/2006 that megatrends are the "great forces in societal development that will very likely affect the future in all areas in the next 10 to15 years". This is the reason why many companies and organizations use megatrends in their strategic work.

"One of the most interesting trends in the past one hundred years is the rise of individualization. This led to the entrepreneurship boom when more and more people wanted to found their own companies and more people were able to do more things as individuals. That's obviously a big promise for the future. We don't need factories; we don't need big companies anymore", says Magnus Lindkvist.

I started diving into the CIFS' data and I found that individualization – in addition to the need to be able to choose everything individually – means that childhood now has more phases. Today, we have toddlers, then young children, after that comes pre-teens (aka tweens) and finally teenagers. The Barbie doll was market leader but because Mattel was slow to note the trend of children rapidly becoming 'small adult consumers', the Bratzdolls quickly took a disproportionately large share of the market leader position. Megatrends can, in other words, be used as an early warning system (aka a weak signal). "It can go terribly wrong when companies fail to pay attention to the development of megatrends. For example, traditional camera and film companies, such as Kodak, are suffering from the rapid rise of digitalization. They saw it coming, but underestimated the speed, with the result that they have had to make massive layoffs in the last couple of years", says CIFS.

Bratzdolls
/ by Dan Burn-Forti

John Naisbitt, whose bestseller *Megatrends* was published 25 years ago, coined the term 'megatrends'. He drew a picture of the future millennium on the basis of ten radical developments thus becoming a pioneer of trend research in business and society.

I took a day off from my usual writing place, the TrendONE Berlin office, and decided to spend it with John Naisbitt and his legendary book at my place in Kreuzberg. No, he wasn't there, but his text was. The premise of the book is that "the most reliable way to anticipate the future is by understanding the present". Then it suddenly hit me: content or trend analysis has its roots in World War II. This method was used back then for obtaining information about the enemies. I remembered and quickly made the connection to my Imageology course at the Communication and Public Relations University in Bucharest. I had to read and analyze letters sent between important historical leaders that described certain people and their behaviors. I had to determine what the most powerful characteristics of that specific person were and build up his profile. This is exactly what trend research and analysis by clustering is all about.

In the same way, Naisbitt started to gather content for the book by analyzing German and Japanese newspapers. In other words: monitoring public behavior. He was the first to introduce this social analysis method into the commercial world. Why was this content analysis efficient? As newspapers have a limited number of pages, they create a mechanical representation of society, sorting out its priorities. So, they edit out what's less important and add the new stuff in. "Our groups collect information about what's going on locally across the country and then look for patterns", says Naisbitt. People, those were the 60s and still the method hasn't changed much – only the medium has been enriched.

Environmental and climate protection, accelerated culture, aging and demographic change are all names of megatrends. You will hear a lot of variants out there: companies give them different titles but essentially they mean the same thing. Also, ranking megatrends between industries is a totally different and complicated story. Different contexts deliver different meanings and therefore, different rankings. Trends as such don't even exist. It is the context that makes them happen. Let's take drinking beer in a Brooklyn beer garden. If people do the same thing in Berlin (and they do!), this will not make it to the headlines because beer gardens have existed for decades. It's just the new location that gives the activity a special aura. Even though megatrends say something in relation to what we already know about the future, it is not certain how society, companies or any of us will react to these forces. "The future is never a given, and anyone of us can affect or create the future. Megatrends have different meanings for different companies, organizations and individuals, because we react, consciously or not, differently to trends", says Larsen in *Why Megatrends Matter*.

A megatrend is something that has been happening for years; you are a part of at least one of them. **Then why do we need companies to point out megatrends if we already know them?** Because even if megatrends will last for years and years, it's important to understand their effects on your business, life and relationships. The sooner you understand it, the more advantages you will have and sometimes it's difficult to distance yourself enough from the concept – sort of not seeing the forest for the trees. Either way, it is happening a lot faster than we think. "One of the first identifiable megatrends is the switch to a sedentary lifestyle by humans in order to ensure their survival. This transformed nomads into villagers, individuals into citizens and marked the starting point of Civilization. It took time to spread across the whole population", publishes the Global Investor Magazine 2.09 in the *In Search of Tomorrow's Global Megatrends* article.

Compare this to the Internet, which in the 1990s took less than ten years to change peoples' lifestyles.
"Currently, it is said that rapid population growth in emerging markets and increasing aging, mainly in developed markets, represent the starting point of the current cycle of megatrends", says Kalbreier in the same article. Megatrends are a tricky thing. Sometimes experts predict that some megatrends will never develop, like alternative energy, only to be proven wrong because they consistently underestimate the power of innovation. A good example is provided by the automotive industry: in 1999, General Motors decided to drop production of the EV1, the first electric production vehicle, judging that electric vehicles would never be profitable. Ten years later, Tesla, an upstart car manufacturer, started the production of its 'Electric Roadster' and recorded its first profit ever due to major innovations in the field of battery technology. Meanwhile, GM has gone bankrupt. This shows the critical importance of innovation and its role in shaping megatrends.

COPENHAGEN AND THE FUTURE

A fat inbox, a pile of unopened letters and (luckily) some Amazon packages are waiting on my desk: normally, returning to the office after two weeks in the sun is the least pleasant moment of the summer. Not for me though, my next thrilling stop is Copenhagen! My friends say it is one of the top cities in the world in terms of quality of life. They also say it's like a live cinema where everybody looks like Barbie and Ken. Since I enjoy sitting down on a public bench and people watching, I suppose it's a perfect destination. However, I am there with a different purpose, meeting Martin Kruse from the Copenhagen Institute for Futures Studies.

Martin Kruse

In 1970, Professor Thorkil Kristensen, former Minister of Finance and Secretary-General of the OECD, spotted the need for Danish companies to have a better view of the future and to be competent in foresight in order to strengthen the basis for strategic decision making, so he and other bright minds formed CIFS. Now, they've gone global, acting mostly in Scandinavian markets but also in the rest of Europe, Africa and Asia.

As megatrends are general pictures of the future, showing us where the world will be in a decade, this information is important to use when creating long-term strategy inside a company. CIFS' focus is on megatrends and other types of trends in terms of sources of innovation, but they are also a consultancy institute so, their core competence is actually strategy. Thus, this would be CIFS' application of megatrends.

// UNDERSTANDING MEGATRENDS

I was excited. I felt like we were starting to talk about the essence of things and I was about to find out the *real* truth. Out of all my interviewers so far, Kruse was the only one who talked about people and how important they are in the innovation process. That impressed me. As I understood it, they use a wide variety of sources for trends, from the OECD, IEA, UN, World Bank, IMF and other big associations or governmental agencies, but CIFS digests the information by themselves. They analyze and interpret. How do they digest trends? "It's about people. You can't just have anyone to do this. The last time we were hiring, we had 140 candidates that fit the formal description, but after the normal test, the candidates were also put through a general knowledge test and only one candidate had the same score as the average score of the futurists employed at CIFS. They need to be 'in-the-know' with many areas and have the ability to create touch-points between different fields of knowledge. To understand megatrends, you need to have a strong comprehension of social science, economics, and technology so that you can work with specialists from these areas."

Kruse said, in terms of methods, "a lot is based on practical experience". That's comforting but I expected a concrete answer about a systematic process and I realized that this doesn't really exist. He continues, "We use a lot of different methods and shift between them based on experience, but we don't talk so much about the method we use. The problem with many of the academic papers where you read about methods is that those methods, in practice, are integrated into an iterative process. It's nice to know them, but in the process, they just mix. It's about knowing when to use what. When we teach people here, we not only teach them the method but we focus a lot on their application. We are more practical oriented than theoretical in our approach".

Again, if we all know about megatrends, why do we need a company to give us this information? There are plenty out there who would say "Ah, I can do it on my own". "No, sorry!" Kruse jumps in with the answer. "It will never work for a number of reasons! Megatrends are simple on the surface, but much more complex underneath. What you don't know is how megatrends will affect your industry. Specific knowledge on the industry, business trends and the touch points between those two are the key answers. Practice makes the difference. A lot of it is knowledge based and not really megatrends related because you can almost 'Wiki' those but it's about how to apply them, it's the experience. Moreover, it's about using megatrends to have a glimpse of the future and this is when the job gets tough". Megatrends are easy when you are looking at how a specific one affected everything in the past but to use them as instruments for 'looking into the future', they need to be narrowed down to specific information, subcategorized into trends, and then you can figure out what aspects will actually influence the industry that you are working in.

NO SUCH THING AS CERTAINTY

But how can you acquire the ability to extract the important info? Or is it like a birth given talent? Sure, you can learn it if you have the motivation and the right teachers, you can do it. "You need not only a broad knowledge base but also creativity and stamina. The reason is that a lot of the barriers that we face are a question of going against the flow. Back in 2007 when all the banks, the IMF and the FED, said that the economy was fundamentally strong, we had a conference with important people from the financial sector. We proposed two scenarios: the main stream (what the central banks and specialists were saying) and the alternative scenario (the economic crisis), which of course they didn't believe in, saying that this crisis couldn't possibly happen. Being able to build an alternative scenario, when everybody says the opposite and almost every data you get tells you that you are wrong, supporting something that is very unlikely to happen, is a very hard job", says Kruse. At a personal level it is very hard as well. Imagine, you need to be able to go against the flow.

I was then wondering: are CEOs too afraid to base their strategies on micro and macrotrends? Are megatrends a safer approach because of the long time span? It turns out not to be about the time span but more about uncertainty. This, my friends, exists in the everyday work of futures studies. So what do you choose: dynamic uncertainty or a mediocre safe approach? "Decision makers and politicians have hard time acting on uncertainty. We have seen this, for example, in relation to political action on climate change. Politicians like certainty but when it comes to the future there is no such thing as certainty and this becomes a problem for decision-making. Being able to take decision based on uncertain information is one of the key characteristics of great leaders. Taking into account what could happen and integrate it into planning is an important part of good planning", says Kruse. We can't focus only on what is most likely to happen, we need to prepare for that which is unlikely but has a high impact as well.

MEGATRENDS AND SCENARIOS

Suddenly, my talk with Kruse switched to the subject of 'scenarios'. Kruse's colleague, Gitte Larsen says in the same article from the institute's magazine Futureorientation that "megatrends are the forces that define our present and future worlds, and the interaction between them is as important as each individual megatrend. That is why futures researchers, companies and others use megatrends when they develop and work with scenarios. Megatrends can be a starting point for analyzing our world".

This is the power of megatrends. Our conversation, both mine with you my reader, and mine with Martin Kruse, has now climbed to a different level. Sure, megatrends are global and there is a huge step from macro to mega; bigger than I had previously thought. At this point it is important to understand the use of megatrends.

Scenarios are, among others, one of the ways to apply megatrends in future studies. There are two different approaches in future studies in terms of the way scenarios are built: the new school method and the old school one. In scenarios, from CIFS' point of view, you can go two ways: you can do complex computer modeling (the use of a computer to develop a mathematical model of a complex system or process and to provide conditions for testing it) or engage in a more intuitive approach. I suppose you've already guessed which one is the new school and what CIFS is doing. If you said they are taking the intuitive approach you are absolutely right! Again, I'm surprised by the exactness of this industry. But what can you do? Keep going, after all our intuition can get us quite far.

In the beginning of the 20th century futurologists thought they could predict the future, their legacy was to create a model and thus create scenarios that depicted the future based on that model.

The **modeling approach** tries to anticipate the future using models build on past experiences and so, it gives a false sense of security and creates blind spots.

The **intuitive approach** fundamentally states that we can't predict the future, but we can fire some arrows into the future-pointing out where we need to focus. Of course, there are all kinds of data and statistics involved here as well. But you don't use complex modeling to decide whether one scenario is more likely to happen than another. Doing so will always get you to a mainstream scenario. Like the one from the economists all over the world who said that the crisis wouldn't happen because it seemed so obvious at that time that it would not. This type of scenario is not interesting from a strategic point of view. It will just tell you what you already know, right?

Instead of computer modeling, CIFS uses expert's interviews and the Delphi method (this is something I will detail later in the book). Although economists sometimes frown upon the intuitive approach, research shows

John Naisbitt once said that "intuition becomes increasingly valuable in the new information society precisely because there is so much data".

that it works just as well and is a lot cheaper. CIFS' job is not to model different scenarios but to intuitively extract information and make sense out of them by asking the following question: How will this change the competitive environment of a particular industry?

The distinction between the intuitive approach (mainly used in social science, political science, etc.) and the modeling approach used by (economists and physics etc.) is a distinction that Karl Popper named as the anti-naturalistic and the pro-naturalistic approach but it is not formally agreed upon.

Ok, ok but I want to know more about megatrends as such. Through my discussion with Kruse I found out that megatrends could be used to extract data. Let's take globalization for example: you can look at statistical data and try to find out if there is more or less of a globalization effect happening in a particular area of the world. This could be part of the complex computer modeling approach (planning or forming a pattern after a model). Modeling would basically mean taking data from the past, watching how the specific trend evolved, how different variables interconnect and extrapolating the same tendency to what's going to happen in the future.

Megatrends are more often placed in the intuitive approach. You take those trends but don't 'insert them into a complex computer model.' At CIFS, they would say, "Globalization is happening and the effects of it will be more labor going to China and India. How will low cost countries affect your company? Are you prepared for lower prices? What will your new value delivery be so you can compensate for loss in revenue?" Don't try this at home! It's best to leave this type of intuitive thinking up to the professionals. Allow me to elaborate on why this is so important.

We see the world changing fast due to technological and natural developments moving at an accelerated speed. From now until the near future, we may not be able to rely on **sustainable innovation** (which means a predictable pattern of innovation that corresponds to the modeling approach) but depend more on **disruptive innovation** (breakthroughs and unpredictable developments that are more in tune with the intuitive approach) which seriously changes the way we look at the world.

"Sustaining innovation involves continuous efficiency gains within the existing production and organizational form. Productivity should grow 2 to 5% a year. Disruptive innovation implies the existence of entirely new ways of doing things. Historic examples are the introduction of assembly line production, the supermarket and McDonald's. Disruptive innovations usually involve not only the core process, such as production, but includes a fundamental change in the value chain", says CIFS.

Kruse gave me an example: fusion energy, power generated by nuclear fusion reactions. Scenario: what if fusion energy hits the mainstream? Applying the scenario to a client in the oil industry leads to the question: What will the prospect of fusion energy mean for oil investment? We know that the general time between finding and extracting oil is 18 years. If we think that fusion energy is coming, then, in those 20 years preceding fusion energy, it might be a good idea not to invest in oil discovery and production anymore. This is an intuitive scenario; a mainstream scenario (aka modeled) would normally not take fusion energy into consideration because it is uncertain.

After all this talk I couldn't help but notice how tightly megatrends and scenarios as methods are tied together. Then I asked Kruse if there are any other applications of megatrends. "Yes. There are. At CIFS we mostly use megatrends as part of a scenario process in strategy or product development, but the use of megatrends are essentially about opening the eyes for the future and in doing so, the process can become a tool in organizational development. The work with megatrends is a learning process for participants and a mental challenge in itself – stimulating the understanding of what the future holds and how we can influence it. This enables better knowledge sharing and new ways of developing ideas, it increases strategic capabilities, reduces reaction time and thus creates a more agile organization with a common framework", says Kruse.

Let's take aging for example, as a megatrend. This suggests that people live longer and it leads to a growing need of products that target the elderly. In a sense, this is not a scenario but at the same time it is helping us to make decisions, or perhaps an assessment of ROI (aka return of investment). It will lead us to the conclusion that if we now invest money in production facilities to create products that target this growing sector, we will have a certain competitive advantage in the future. So, in this way you can use megatrends as something other than scenarios but normally you would combine this information with scenarios as well. They go hand in hand. Transforming the information into a scenario and adding some uncertainty into it would look like this: We know that elderly people are hard to persuade. The way they purchase products was created a long time ago and they are used to spending smaller amounts of money. So, even if there are relatively more elderly people, we can't be confident that they will spend more money. You don't always need the scenario itself but it is a good way of testing your strategic decisions.

RIDE LIKE THE WIND

I left Copenhagen with a good feeling about how my findings have developed. As I was waiting in the check in line at the airport I tossed around the topic in my head. Today, megatrends are a relevant strategic subject in many corporate offices. Some megatrends such as demography, health and mobility are on the strategic agenda. In practice, however, each company has its own perspective. **But how do megatrends add value for businesses? They can only add value if they are intelligently translated into a specific business context.** They are very effective tools in the future areas of innovation, markets and products. As I mentioned before, it's not the trend that is interesting in itself, but its strategic implications.

Naisbitt writes in *Megatrends* that "trends, like horses, are easier to ride in the direction they are already going". Those who ride global trends find excellent opportunities, while those that run against them tend to find that the going gets very difficult very fast. Identifying trends and identifying which companies will prevail in those trends, seems to be vital to long-term success.

Overall, we had a thought-provoking look at megatrends here. Copenhagen is just a stone's throw from Hamburg so that will be our next stop – to learn how to read consumers' minds.

But first, let's see what a megatrend looks like in a briefing format.

// NETWORK ORGANIZING

Out of the ten megatrends that were republished on CIFS' website in a shortened, edited form, developed by Kaare Stamer Andresen, Martin Kruse, Henrik Persson, Klaus Æ. Mogensen and Troels Theill Eriksen, from Copenhagen Institute for Futures Studies, I chose to publish the megatrend related to Network Organizing.

"To enter a network is a natural part of being human. Central to all networks is communication, because communication is the reason we have a society, a culture, an identity and an economy. Network organizing is a megatrend because network has become a central term that permeates our way of thinking. Cheaper transport, better infrastructures, the Internet, mobile telephony and increasing prosperity have revolutionized the opportunities for communication and network organizing. This megatrend is, in other words, closely connected to the development in several other megatrends, not least digitalization, globalization, and individualization, but also prosperity, immaterialization, and commercialization.

A network's value increases exponentially with the number of members who are in it. Changes in a network society do not happen linearly as they do in an industrial society. That means that many changes that took decades in the past now happen significantly faster. An example: just two years after the World Wide Web was launched in 1992, 10 million users were on it, while it took the telephone four decades to attract the same number of users. Network organizing greatly affects technological, societal, and economic development, and we have probably seen only the beginning. The rapid development potential in the network society means, on the one hand that companies can expand incredibly fast, as happened with Microsoft, but, on the other hand, companies in all industries can risk out competing each other in a very short time. This applies even to Microsoft, which, even though 90% of computers use its programs, is losing share to the free operating system Linux.

Networks drive out hierarchies and create many new open and decentral social structures. This applies to private life, especially for the younger generation, to the labor market, and business life. Medicon Valley in the Øresundregion is an example of one of Europe's largest clusters of biotechnology companies.

Network organizes also promotes urbanization, because urban regions with good infrastructure, good development possibilities, and a rich research environment attract the creative class.
Google is an example of a company where the network principle has shown itself to be a good business ideology. The Google search engine's strength is, in fact, that it lists search results according to how centrally a web site is in the network – that is, according to how interesting users believe it is".

Trend, trends, trends. Are you still with me after so many concepts? Ok, let's revise a bit. Megatrends differ from other trends in terms of:

> **Time horizon** – they are observed over a period of decades;
> **Range** – they are comprehensive and extend to all regions of the world;
> **Strength** – they are broad and profound, affecting all actors - governments, individuals and their consumption behavior.
They also affect companies and their strategies.

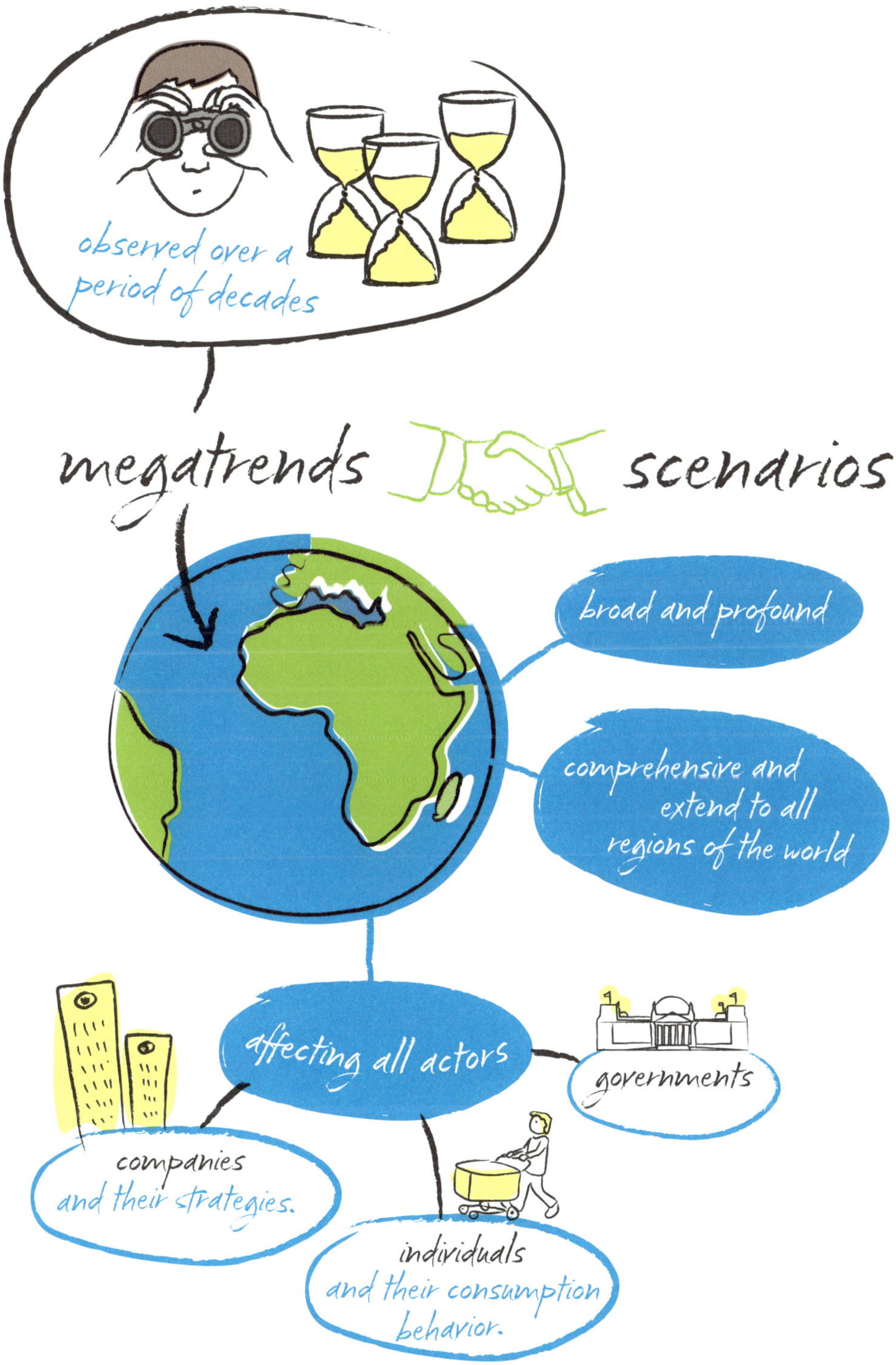
observed over a period of decades
megatrends
scenarios
broad and profound
comprehensive and extend to all regions of the world
affecting all actors
governments
companies and their strategies.
individuals and their consumption behavior.

CONSUMER INSIGHTS

Back in university, when I was taught how to develop concepts for communicating a brand, I was told, among others, to ask myself some basic questions: If your brand were a person, would it be female or male? What eye and hair color would it have? Would it be phlegmatic or more of sanguine? What kind of story would it tell? What would be its favorite fairy tale? Would it be more of a beer or a wine drinker? And why? This brainstorming technique probably wasn't called a consumer insights method but it had a few things in common with it. When you have to place a brand in a conversation it's much easier to develop ideas when you think of the brand as a person. And this is not completely nuts. Nowadays, most consumers associate brands with human experiences and emotions. "It's no longer about what the brand can do for you but about what you can do *with* the brand", as trendwatching.com says. As if it was your friend.

I don't believe in a cookie-cutter approach. But who does? It's quite clear that consumers' values change fast and along with them, their buying behavior changes. Today, people buy products not only for pragmatic reasons but, also for emotional ones. It happens to me all the time.

Brands already know all about this. They've realized that they need a comprehensible vision of what the consumer needs. Imagine if Henry Ford had asked people what they wanted, they would have said "faster horses". So, this is why asking consumers as a standalone method is not a good idea but paired with trends and market insights, they make a hell of a good team.

Insights are unrevealed emotions. Finding insights means:

> grasping the inner nature of things intuitively
> having a clear and deep perception of the situation
> understanding a sudden and complex situation

It's always tricky to quote Wikipedia but I found this person called Conroy that has an interesting statement about consumer insights. According to him or her (as I didn't find any additional information about the person)

an insight "is a statement based on a deep understanding of consumers' attitudes and beliefs, which connect at an emotional level, provoking a clear response (This brand understands me! That is exactly how I feel! Even if they have never thought about it quite like that) which, when leveraged, has the power to change consumer behavior".
Mohan Sawhney, a management consultant, author, McCormick Tribune Professor of Technology at the Kellogg School of Management in USA and the *guru of consumer insights*, defines consumer insights as being "a fresh and not-yet obvious understanding of customer beliefs, values, habits, desires, motives, emotions or needs that can become the basis for a competitive advantage. They often come from unusual sources, are discovered accidentally as they often are rooted in observed anomalies and rarely emerge from quantitative analysis". Sawhney adds that "customer insights is a bottom-up approach that leverages insights into the behaviors, perceptions and needs of current and potential customers by involving them as true partners in the innovation process". Most of the traditional approaches do not seek out or reveal insights beyond those that a customer is able to verbalize. It is one thing to invite consumers to talk about their needs and wants – it is quite another if you seek the unarticulated.
"**Breakthrough innovation can only begin when an organization is no longer satisfied with what customers actually say – these organizations open the door to extraordinary levels of innovation**", says Sawhney.
We are measuring the quality of our lives in social terms before economic ones. Buying by quantity or shopping as a sport is passé. The amount of things available on the shelves has shut down people's mind and now they can only be opened by experiences and social connections. Businesses that recognize this cultural shift and provide unique experiences at the point of sale, environments that foster special moments and unexpected situations, will thrive. But how do we produce these experiences? It's such a difficult task that only a few are able to do it right.
In a 2007 article entitled *How Consumer Conversation Will Transform Business*, PriceWaterhouseCoopers spoke about how consumer conversation will

Engaging with consumers in a dialogue and being an effective listener is where the greatest innovations will come from... **Consumers are in the midst of a conversation that isn't ours**. The race is on to grow ears to learn what they are saying.
/ John Hayes, CMO American Express.

transform businesses. The readers were asked to do an imagination exercise, it went like this: "Imagine it is seven o'clock on a Monday morning. You turn on your computer. On the screen is a startling level of insight into the real-time attitudes, behaviors, and intentions of people. By knowing *why* they buy, you know where markets are going. This insight informs every decision you make as an organization. Every hour of every day your customers help you innovate well in advance of demand. This Monday morning is coming. In billions of chats, blogs, emails, phone calls, and social networks, consumers are talking about brands. Somewhere in this massive conversation are shouts and whispers that will determine their success or failure."

In a world where consumers, before buying a product, put more value on other consumers' online feedback than on any type of advertising; it is clear that the rules of the game have changed. They now expect rapid responses to their immediate desires, and they have the power to demand what they want. But how do you decide which voices to listen to?

Focus groups are a thing of the past; today you have to draw people out in their own environment. / Carla Hendra, Co-Chief Executive Officer, Ogilvy North America.

"In order to reliably identify what consumers want, companies must connect conversations about their own products with market and social trends. Embedded in conversation is what consumers really want, need, and expect from a company", says the same article from PriceWaterhouseCoopers. Careful listening will give you the solutions. I knew a few things about the consumer research methods: surveys, focus groups, cultural deep dives, in home ethnographies, roundtable interviews, scene or trend immersion, shop a longs, blogs, videos, creative journals. **There are various definitions of consumer insights out there but what I was looking to understand was more about insights as an innovation input.**

// @STURM UND DRANG

Trapped in a never ending but pleasant cycle of friends that came to visit me in summery Berlin, I felt the urge to escape. Sure, I was going to the office everyday but sometimes one has to be alone. Yes, alone. So, I went outside to get some air and hopefully discover something new in Berlin. To tell you the real truth, I was in the search for a cozy place to sit in the sun and eat a cheesecake. In less than ten minutes, mission accomplished! I was sitting at a terrace café reading Scenario, the quarterly magazine of Copenhagen Institute for Future Studies and pondering my meeting set for the following day with Stefan Baumann, CEO and Founder of Sturm und Drang, in the MetaHaus building in Berlin.

Stefan Baumann

Later, I found out that in late 1760s Germany, the term 'Sturm und Drang' (which literally translates as 'Storm and Urge'), was associated with literature or music aimed at frightening the audience or imbuing them with extremes of emotion. Goethe was one of its activists.

Sturm und Drang is a qualitative research boutique helping customers to develop innovations and visions for brands based on a focus of consumer and market trends. Their special insight and foresight research inspires brands in their invention and renewal processes. It is Sturm und Drang's mission to translate relevant insights into innovative products, brands and services. "We create innovation for markets that do not exist yet. When we started the company, we thought it is necessary to have future intelligence combined with innovation thinking because in most of the companies you only think inside-out. This means you are developing a kind of technology and try to market it afterwards. We do it the other way around. We are thinking outside-in. Try to do it in a strategic way. That means we are detecting white spots, opportunities, gaps and then we try to define which would be the right product, brand, service that fills that gap", says Baumann.

Going deeply into innovation practices, seems to be a new niche approach that is growing at the moment and providing an interesting twist to their trend research backgrounds. They now use trends and insights as raw materials for innovation, first detecting them and then transforming them into innovation.

Different from the dedicated trend research companies, Sturm und Drang is not selling trends in quantity; they try to define which trends would be suitable for the kind of project and client they are working with. Their main exercise is to sort out trends, meaning clustering and detecting patterns, trying to connect and find the interplay between those trends. Ring a bell? They try to figure out which are the most relevant insights, foresights, innovation concepts and visions. Ok, we are again into the 'precious words' area. I kindly ask Baumann to explain the differences.

Insight is about what's happening in the space between the consumer and the product. It means what's going on inside the consumer, the key motives, drivers and barriers that you as a product developer have to face in the innovation process. "We've defined insights like an inner conflict inside the consumer. It is a very good springboard for us to surf or to provide the consumer with the kind of product that solves a conflict. So first we have to search for insights and afterwards we can go into innovation processes", says Baumann. This is one approach, from insights to innovation and the second is from foresights to innovation.

Foresights are future states, very close to trends. Actually it is just another way of calling trends. I clearly understand why they chose not to use the name trends, because it is often misunderstood. Even I avoid saying it and that's pretty hard when people ask me what my job is. But telling them I am a foresights analyst won't solve the problem. The word 'trend' is overused, mostly related to fashion. It is too close to fads and hypes so 'foresights' sounds more strategic. At Sturm und Drang, foresights mean looking into the collective mind at the societal level, in markets at large, asking what are the interconnections and the context around the consumer.

Why did I choose to talk about all this and not go straight into our topic – consumer insights? Well, because foresights are very strong and influential factors for consumer behavior, coming in the shape of a longing that derives from the foresights. If we spot a place, a pub for instance, with a strong collective wish for regional products, say Guinness, this could happen for several reasons. It could be that people feel 'a longing' for home (home sickness) and that's the reason why this kind of product sells more (at least in areas where there are lots of Irish people). Sturm und Drang tries to understand why and to detect what the interconnections are between the different elements that influence consumer behavior. "Detecting consumer trends that are linked to the crisis mindsets could be one of our 'homeworks'. Our level of understanding society and the world is by examining consumer attitudes and behavior. This is our gate to market understanding", says Baumann.

When it comes to trends, Sturm und Drang always makes a differentiation between analysis levels. Therefore, as a reinterpretation of the already classic micro, macro and mega categories, they spot: **societal trends** (megatrends, technological change, economical change), **consumer trends** (macrotrends – 'creacipation' is one example and it means to be part of a process where consumers want to and actively *do* participate in the product or service creation processes) and **industry trends** (market reactions, how do industries react to a specific event, what kind of product/services they offer and launch. On a very practical level these could be called microtrends).

MAKING THINGS HAPPEN

The consumer trend level is not a specific category but an overall development or shift. "In consumer insights we always think about insights that are specifically derived from a product-consumer relation", says Baumann. For example, let's take drinking beer again. Remember I'm in Germany? Let's say you wish to understand why people do or don't drink beer. A first step would be finding a consumer insights process (e.g. group discussions, observation) and explore it, understanding things on a specific category level. If you want to understand what the consumers' overall longing is you would have to investigate not only beer but also the whole food industry. This is the step where you identify insights and innovation fields. Media screening (videos, blogs etc.) would take us to trends aka foresights and not to consumer insights. Sturm und Drang does both.

The ideation process is the next step after identifying insights and foresights.

What is interesting is that the clients are part of the ideation process. This is how it should always be. The client shouldn't stay at home waiting for the brilliant idea to be delivered to him; he should be part of the process so he can provide valuable input in the form of company insights. "In the process, we try to mix them as much as possible in terms of what departments they come from. And then mix them with lead users", says Baumann.

The ideation process is actually a three day camp run by Sturm und Drang, like a brain camp. The first day is the 'strategic framing', where the client's innovation team and Sturm und Drang's team try to enrich their search fields. This is achieved by talking. Simple conversation is so much more important and productive than you would think. You'd ask: 'how do they do it!? How do these companies and people develop ideas and give names to trends?' Well, by sitting down and tossing ideas around with one another until the information makes sense. Once some possible insights have been identified, they should be objectively and qualitatively tested to determine if they are truly bold or just rephrased – old wine in new bottles. This needs to be done before the process of writing new concepts begins, as there is a completely different set of questions necessary to identify the potential of an insight. What are the questions? That's what I needed to know! Baumann explains that the driving forces, so called 'tensions' which are described within the insight are not conscious enough to be tested verbally. The insight works at an unconscious level. The proof of the insight is the concept (product, service, communication) that works for it.

"To assess insight potential we always ask ourselves 3 questions:

1. Does it start with a true and interesting phenomena observation?
2. Does it incorporate an underlying motivator (fear, desire) and a tension that is caused by an apparent contradiction?
3. Does it trigger action?

The more you say yes to all of these three evaluation questions, the more energetic the insight appears to be; the better it is to work with", says Baumann.

The second day of camp is the lead users day when they are asked to give feedback and brainstorm on different creative techniques in a moderated process. It's all about brainstorming techniques. One of them is called speed dating. It's a funny thing: taking the idea of sitting face to face and talking to a potential partner for a short period of time, to the business world. The third day is about the process of bringing ideas into concepts. How do you do that? You work with illustrators, write, do paper prototypes, sketch etc. Then, one of the concepts goes into quantitative tests. Consumers have to score, rate (the classic way) how much they like it or how relevant the product or service is. "A truly good insight has the ability to be executed in many different ways. If you can develop only one possible concept from an insight, it is simply not an insight. Insights allow multiple executions, so when you have identified all of these directions, you can develop the concept executions and scan them to find the most compelling and unique delivery of your insight", says Baumann. For example, the basic formula for beauty can have various meanings in different cultural contexts.

The cultural context of beauty

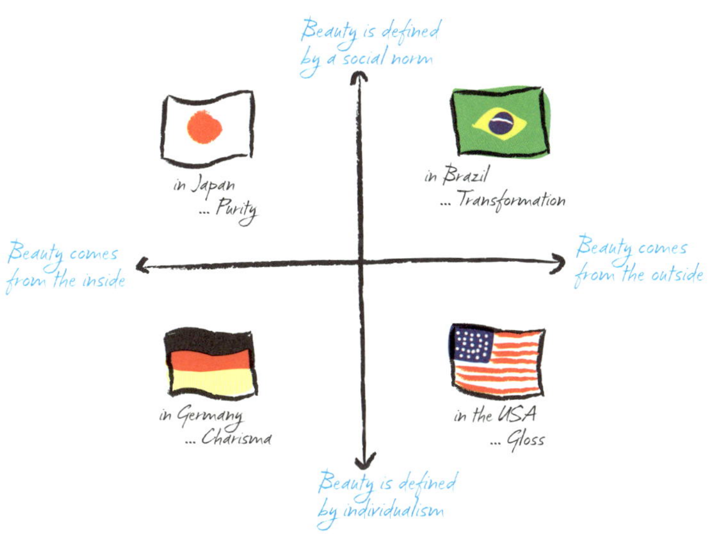

Here is how Sturm und Drang interpreted beauty in the study *German Beauty – How Cultural Codes Influence our Understanding of Beauty* in December 2009.

Trend research is tightly linked with semiotics analysis because one has to understand the key motifs of the society when spotting a phenomena and explaining why is it happening. "I am a psychologist and this means I need to understand what drives people to do certain things and I need to understand 'the why' of everything. Why is the culture behaving like that? What's the reason behind it? I am always interested in the consumer and the culture. I believe they are deeply linked", says Baumann.

// FUTURE OF CONSUMER INSIGHTS

I couldn't keep from asking what the future of the consumer insights approach to innovation will be. According to Baumann, this is the answer: "We are working on different digital platforms programmed into the knowledge management systems of the companies. We provide a tailored database and install the system in the companies, updating it quarterly. Trend management is like a service process where you can subscribe to quarterly update of your trend feed. And you have to work with that knowledge. That's where we come in with our services. I think that's the future in our trend management processes and it goes together with the innovation management as well. So, extending the online tool. **The age of trendreports is gone.** There's no future for it. It will be replaced with a data fuel, which is directly uploaded to your system, even filtered. Like TrendWiki. [Remember we talked about TrendWiki in the weak signals journey to Helsinki?] The fact that companies work with trends, rate them, give comments, use them as living currency, is a very interesting development".

So, guys, for a good product you need the right timing and the right insight. As a company, Sturm und Drang is valuable because they are translators of consumer longings, specifically those which customers are not able to verbalize. So, their job is to interpret their motifs. After this, within a systematic process, Sturm und Drang provides ideas that are answers to these consumer's longings. By doing so, they create markets and categories that don't exist now. "This is valuable for saturated markets. I don't think one can have valuable innovations these days if he just listens to what consumers say. One has to understand them better than they do by themselves. By doing so we ease the processes of our clients, help them to organize the innovations from outside in", says Baumann.

Consumer insights is a mixture of gut feelings and strategy.

At the beginning of the process you have to be calculated and visionary and at the end you have to be very creative. Therefore we shouldn't forget to speak about creative impulses in the research process as well. If you are not creative there, you are not getting anything from the process. Because consumer insights are not about data crunching, they are about speculating the future and involving people – the human factor in that process is most valuable.

At the end of my interviews I always ask for tips on further readings. This time, I was surprised. I was told: "I've got some books on my bookshelves but I didn't read them. I think consumer insights is such a fresh discipline that I have to experience it by myself. It is not so much about theory. There is no clear definition of consumer insights. You have to find yourself a good one", says Baumann. They sure have.

The famous marketing guru, Philip Kotler, once said that "consumer insights is the new paradigm of marketing science". Clearly, a consumer insight can reveal a symbolic and emotional bond with a brand. But it's hard to get an insight. It comes after a long work of observation, thinking and immersion. Traditional research is no longer innovative. That's pretty obvious. Humble researchers have been replaced by insights managers. Products are branded based on insights and innovation is driven by insight.

Consumer insights are like icebergs. The really interesting part is below the water line. What you clearly see with your eyes can either be misleading, or of not enough substance to aid in developing future strategies. Remember that consumers, not marketers, lead the discussion.

It's time to cross the Atlantic – on to The Big Apple. Yes, New York. New York. But before this, flip the page to check out the cider story!

// THE CIDER STORY

It is always a struggle to publish case studies in this industry, as companies are afraid somebody will steal their future strategy but anyway I managed to stumble upon a website with some old ones. Why am I writing about them if they are old? Because they are still very good examples of how consumer insights changed brands. Context-Based Research Group was established in 1999. The company combines the qualitative research skills of cultural anthropologists and the communications and business strategy of marketing experts. On their website I found this cider story.

> "In early 2001, Boston advertising agency Clarke Goward was pitching the makers of a hard cider beverage. Knowing that they needed an edge to win the business, Clarke Goward worked with Context Research to gain insight into the world of hard cider drinking in the US. They needed insights to fuel great creative work and materials that would help them present – and they needed them quickly. Context anthropologists visited bars in three cities across the US to learn why people choose cider and what happens when they drink it. During their time in the pubs, the anthropologists conducted participant observation and interviewed a total of thirty cider drinkers. Based on these sessions, Context delivered insights to fuel the Clarke Goward creative, and great quotes and photos the agency used to present their strategic thinking. And, the entire project was done in just a few short weeks. Chris Lawson, Executive Vice President at Clarke Goward, describes the result of working with Context on this project:
> 'The New Biz ETA worked extremely well as part of our pitch. It helped the client see real customer reactions to their product. It got their attention, enhanced our creative, and helped us win the business.'"

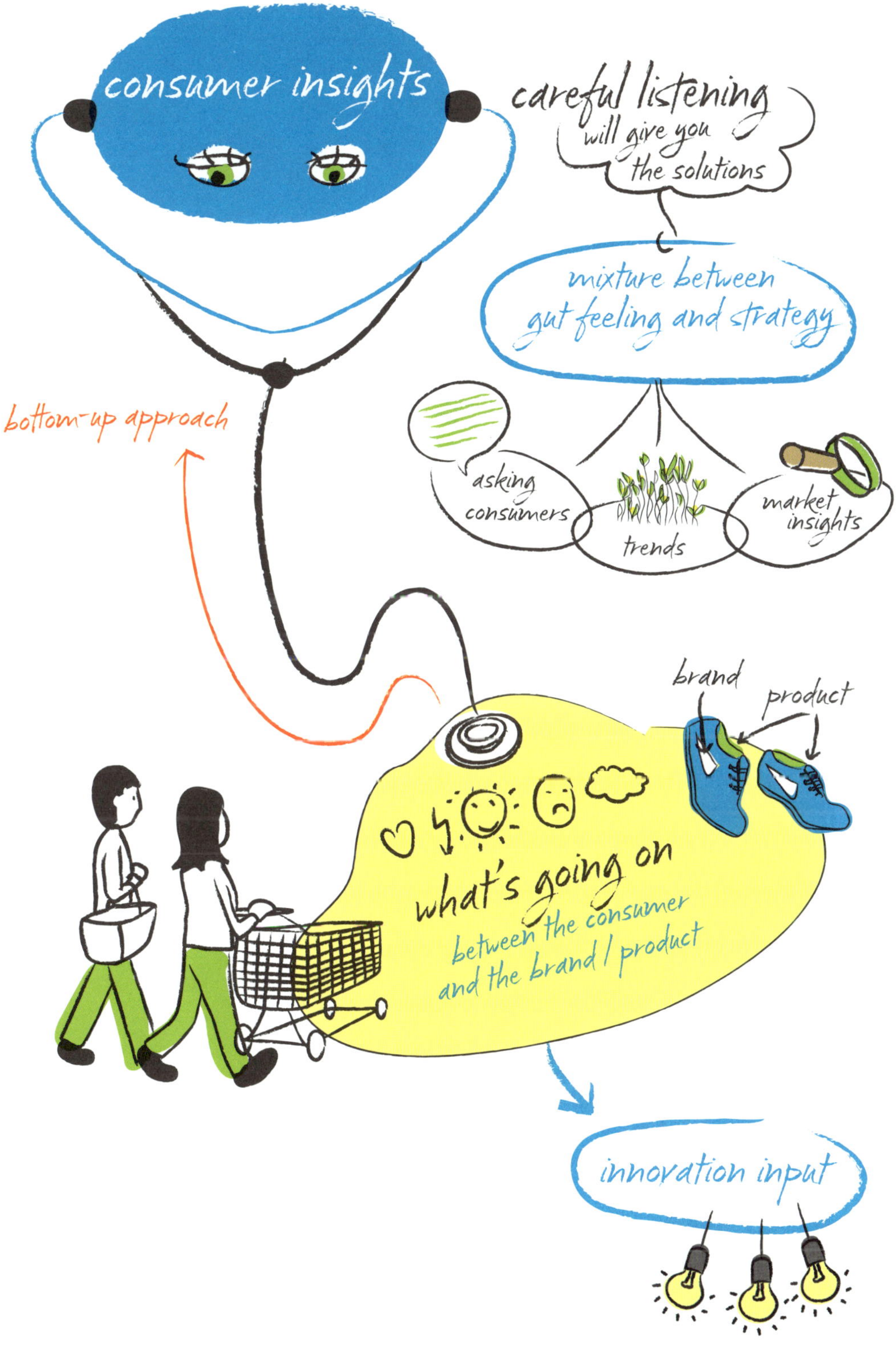
consumer insights
careful listening will give you the solutions
mixture between gut feeling and strategy
asking consumers
trends
market insights
bottom-up approach
brand
product
what's going on
between the consumer and the brand / product
innovation input

GOT THE TRENDS. WHAT NOW?

Ideation

kr: Justin Rothbart

TREND CONSULTING

I'm the kind of person who can work until midnight, but doesn't function well in the early hours. It was Tuesday morning and I was in the kitchen of the TrendONE Berlin office, preparing my morning coffee and trying to prevent my eyes from shutting, when I greeted the person who was sitting at the communal kitchen table, having a conversation in German with one of my colleagues. As it was morning and German is still not my strong suit, I paid attention only to my coffee. All of a sudden, Norbert, my colleague starts speaking in English saying "Delia is from Romania and she is writing a book". He then adds, indicating to his companion, "Florian has a company in New York called CScout". My mind reacted: no way!? I couldn't believe this. Rewind. I was talking to Florian Peter for over an hour on the phone a few months before this happened. But as we never saw each other's faces we just didn't realize we were in each other's presence. He said: "well, I imagined a Romanian looking differently. You look more Italian or Spanish". Honestly, I get that a lot. Why do people think Romanians are actually blond and very pale-skinned when most of them have dark skin and dark hair color too? I mean, yes, we are close to Poland but we are more Latin than anything else, both in language and in culture. So, anyway, I just butted in on Norbert's meeting and started talking to Peter again about the subject that I was interested in for the book – consultancy based on trends. As I mentioned above, I spoke to Peter, founder of CScout in New York, at the very beginning of this project. The reason I am writing about them now, here at the contextualizing step of our journey, is because they blend activities.

Florian Peter

CScout is a kind of piece that fits in more than one place in the puzzle, due to its use of an interesting combination of methods. They work well with other trend research companies in joint workshops; they do trend consulting and juxtapose their trend findings with consumer insights. They are inspired by companies like IDEO – a design and innovation consulting firm and Doblin – an innovation consultancy company.

Peter's initial idea to start CScout is, in my opinion, one of those success story examples with a perfect sequence of happenings. He was working in a multimedia agency in Berlin in the mid-1990s where he noticed that much of the cool information and inspiration they were receiving came from the United States. There was an information gap between the US and European markets and he concluded that there was a need for an information broker. So, Peter went to New York and started, with a couple of friends, a magazine called *Flyer* that focused on trends and lifestyle – a kind of culture magazine for the youth.

FROM A CULTURE MAGAZINE TO TREND CONSULTING

Flyer was launched in Berlin by Florian Peter and his friend Mark Wohlrabe in the mid 90s. The concept grew and hit the States in 1998, continuing to run until 2002 when CScout was born. How a lifestyle magazine is related to trend consulting and whether the research technique stayed the same is what I wanted to find out. "Essentially, if you are a small magazine, you always talk about trends with your advertisers. Back then, they were clients like Diesel or Coca-Cola. We were telling them about our readers and what they like, why we cover the stories that we are covering, what is cool and what not", says Peter. In a way, without naming it, *Flyer* was composed of a trend research team spread all over the world that helped brands get into an early adopter/hip-minded type of audience. With an international network of writers in New York, Berlin and Tokyo, Peter received global trend information in arts and culture and was communicating it to potential advertisers.

The year 2001 had a huge impact both in history and business – 9/11, the recession and the '.com bubble'. Magazine advertising dried up and Peter was forced to turn to something else and the move toward trend consulting came naturally: "I used the knowledge of how to deal with brands and communicate culture trends and the knowledge of operating an international magazine with an international network of collaborators to turn to a consulting business. The move from print to conducting research, publishing trendreports and doing presentations felt natural and so, CScout was born".

// TREND CONSULTING COMPONENTS

How does trend consulting work – what does it even mean? CScout is not only talking about megatrends like 'urbanization' or 'aging' but also about more explicit trends (macro and micro) and their business applications. They are not just saying 'here's a cool way of doing augmented reality' but they are deciphering what that means in a specific marketing context and clustering microtrends into macrotrends. "It is difficult to categorize. People question trends very fast, asking what you mean by, for example, 'women social networks.' It is definitely not a microtrend but not a megatrend either. As a company, you have to define it, give it a name and work with it in a meaningful way – you have to be able convert the abstract trend into an applicable business strategy", says Peter.

THE TREND SCAN

Scouting is one of CScout's main research methods. Trend scouts are activated to deliver fresh insights on the specific trends, categories and client industry. The scouts deliver macro and more importantly microtrends from the specific markets.

Sounds to me like the research technique hasn't essentially changed, even though the transition from a magazine to a consultancy company has been made. For a magazine, the authors are briefed and assigned a topic, given a word count, a budget and told to hit the road. Now, in the trend research world it's similar: "We put a consultant or a trend scout on a topic (for example, food trends in China) and we tell the person in China to research for interesting related trends. The way he/she communicates is by writing an article: might be brief or blog-like writing style or case study driven but it's a similar process", explains Peter.

The truth is trend scouts seldom hit the streets anymore. Instead they participate in the selection of innovations by constantly scanning, with a discerning eye, blogs and websites in their local culture and language. They are experts in finding new things, looking at what people post and commenting by posting themselves. **Want to know why they don't do so much street investigation?** The scouts look for inspiration before it hits the streets, by talking to innovators, entrepreneurs, going to conferences and industry fairs. Physical street-scouting makes up only 10-15% of CScout's scouting. Innovation should be spotted from its earliest point in life, even before it gets incorporated into brands and street elements (guerilla campaigns, shop windows, outdoors). On the other hand there are plenty of ways to do innovative things with the technologies that already exist and get inspired by the minutiae of everyday life. So, street hunting is not to be trashed.

CScout is proud of the worldwide network of scouts they work with. And they have all the reasons, it's an authentic one. If let's say a sport manufacturer comes to CScout and asks for wearable trends in India they actually turn to an Indian collaborator who will query his trend network, and conduct research in local media and search engines.

Peter gave me a simple example to understand why **local information matters** so much: "You really need to be in the market where the information comes from. This way, the information makes sense to you. For example, if you see a campaign coming from Burger King Brazil, you should be getting this information from a Brazilian based in Brazil because only they can describe and know exactly what this means." Local scouts rule! The language is also important. The culture gives a totally different feeling to the written text. It's like reading translated poetry. You know it is never going to have precisely the same meaning.

To further enhance the trend scan, CScout adds expert, consumer interviews and ethnographies to build topic-focused trendreports. They take the inspiration that comes from the trend scouts and use their internal trend archive, which they refer to as the 'trendpool', aggregating the relevant trends for the project they are working on.

You probably wonder, besides trend scouts where does the information come from? They buy reports from other research companies; I noticed many companies do so. This is actually very cool. **Acquiring the missing information needed from another company allows both companies to be less competitive and expand their knowledge thus what they are able to offer.** This act of acquiring/sharing needs only a name and a framework and it could become an industry in itself, right?

IDEATION WORKSHOPS

Trends are condensed for a client through a strategic consulting project. Again, one of the keys is tossing the researched information (the trend scan) between co-workers and other experts and even non-experts on the topic, trying to develop what makes sense for the client. You don't just show your findings and say, 'do with this what you will' but actually come up with a strategic analysis. Sort of adding, connecting the dots and giving recommendations.

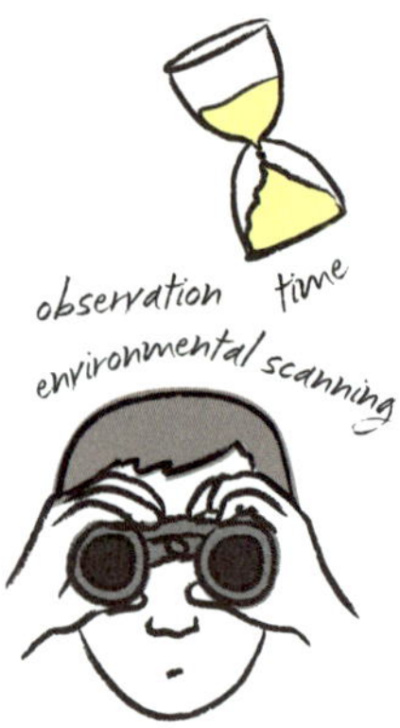

To extract the meaning out of a trend, you need to spend some time exploring how it might evolve over time. In fact, insights need to be spotted during the scanning process and then you can see how they connect or operate in isolation from one another. There could be weak or strong connections between trends, which can lead to an interesting new story and at the same time – some trends that seem powerful might collide.

To develop actionable ideas based on trend research, CScout conducts one - two day ideation workshops. Besides the client, they invite an innovation panel of experts, power-users and utilize innovative idea-creation exercises (eg. storyboarding, mind mapping). The strategic ideation workshops can convert trends into innovations, guiding teams to apply research from the trend scans to business objectives.

There are various add-ons used to enhance the ideation workshops, like **trend immersions**. Also called trend tours, they are especially popular in Tokyo, New York and San Francisco and they can be a good starter for ideation workshop, as it immerses the participants in a specific trend by making them experiencing it while on a tour through the city.

CONSULTING

At this step, based on the output of the ideation workshops along with the client, CScout details the actionable ideas, getting them ready for prototyping or implementation. Often, the consulting part is about checking if the idea will work with the company, whether it has potential and can be executed. This is the time for giving the idea business potential by conducting a detailed

analysis where legal and financial aspects are also taken into consideration. This part is co-created together with the client.
The ideas come out of the workshop in a broad form, eg: develop a mobile Augmented Reality game. But what does this really mean for the client? (It could mean for example, doing a scavenger hunt to unlock special deals by viewing the augmented information attached to billboards) What agency will work together with the client to implement the idea and how much would it cost? Does the idea work with the brand's culture? (Maybe in the workshop people were very excited and creative but the idea doesn't fit.) All these questions need to be answered. In order to do so, the steps of the trend consulting process would ideally be followed as explained above.

IDEA EVALUATION

Of course, the criterion for the evaluation depends on the type of company and product. But what I found interesting (also noted at the beginning of this chapter) is CScout's blend of methods. **They use consumer insights to juxtapose with trends.** "Consumer insights is a very strong field. We like it, it's powerful. The microtrend is the idea and consumer insights comes in for testing the idea", says Peter. I wondered if there was a special consumer insights method that you could use only for juxtaposing with trends. According to Peter there are at least two ways. You can start from a hypothesis microtrend like: I think people are going to watch everything in 3D in the near future. Then you test it with innovative consumer insight techniques (we talked about some of them in the previous chapter so please fill in what you know here, e.g. consumer observations) in order to find out what people think now, what they might think in the future and what they can't even imagine thinking. The other direction would be the 'tabula rasa.' This means absorbing as many ideas as possible from an innovation panel (selected group of experts, academics, power-users such as bloggers and outside thinkers, not connected to the topic or industry) and innovation fields, discussing macrotrends together with them and then filtering this new information in order to discover the smart ideas that lay beneath the surface.

// THE FUTURE OF TREND CONSULTING

"The future is already here; it's just not evenly distributed." This is a famous quote by William Gibson, and one of the most popular sayings in the futures industry. Here's an example of why: Let's say you work for Audi and find out from your trend scouts, your trend research agency or by yourself that an unknown start-up in Australia has come up with an innovative idea for more efficient car parking sensors. For some time now, your team has been working on a way to make Audi, as a car company, more functional by using things such as embedded digital technology so its cars can be parked more easily.

The start-up from Australia has delivered the solution or innovated right now. So, the future is already here. But still, you should see this as a strategic opportunity for Audi. The future is not widely distributed as in communicated or strategically consulted yet. The trend scout reports on the Australian startup, so he is basically distributing that piece of information abroad and distributing the future by communicating the information further for others to pick it up and transform it into something new. Therefore, the future doesn't have to come out of nowhere. **The trend scout enables connections by distributing knowledge**. This is the role of a trend scout whereas the role of a trend consultant is to take these pieces of information that appear in the form of microtrends (in this case: the innovative car parking system) and put them in a context, creating a new innovative idea (e.g. embedding parts of the discovered technology in the car's dashboard and expanding on it by adding a twittering system for parking lots that tell your car which is the nearest available place where you can park). "CScout comes up with new ideas in a case like the one above by thinking of the future in a hypothetical way. Meaning, you assume certain scenarios like: everybody moves to a big city and you can't drive anymore because there are too many cars parked", says Peter.

The general public is familiar with big consulting companies' structure and methodologies. A typical one would sound like '10 steps approach to innovation' or '7 pyramid approach'. In the future, companies like CScout that have less rigid approaches, should pop up more and more.

The market is going to be more dynamic, more in need of innovation but at the same time, more competitive. "Every company we work with now has an innovation manager. This wasn't the case 6 years ago. But innovation, for the companies, is not about coming up with a better way to produce *the* butter but it's about the team that looks at all the tastes, ingredients, values consumers attribute to butter in general. Then come the strategic questions like should we do it fair trade or having the LOHAS in mind?" says Peter. Companies started building these innovation departments in hopes of coming up with new products. Various case studies have shown that this is not enough in order to have a global view on how external elements affect an industry. Usually, you know your company's work very well but are less informed about external factors that can affect your industry. This is why collaborations between an internal innovation team and an external agency work best.

In this pool full of innovations there are a lot of people swimming. I couldn't help but wonder, is it more common for consumers to innovate or for companies and brands to do so? I guess it depends on many perspectives. As Peter mentioned in our talk and I am paraphrasing him here, the abundance of choice out there opens peoples' appetites and makes them want a new type of coffee every 6 months and not every 2 years. It is globalization that leads to more innovation because it allows brands to travel all over the world without much effort. You want the coffee beans from a fair-trade farm in South America? No problem! In addition to this, there is the constant consumer need to feel thrilled and live new experiences. "They are not happy anymore with an ordinary piece of bread and butter. It has to be a super cool kind of bread and a super-cool kind of butter. People are much more demanding, much more fragmented in terms of likes and dislikes because society gives them the possibility to choose a product that suits their lifestyle 100%", says Peter. And they are determined to look for it until they find it.

Consumers want to have choices, so companies have adjusted to that.
We have crossed from finding out where ideas come from to how we can utilize them further. Now we have a broad knowledge on how trends emerge and how to grasp their meaning. I think at this stage it is important to recognize three things:

> **Trends don't exist in isolation.** They have to be placed in contexts so they can really be useful.
> **To understand trends one has to know about the past.** Trends have connections with the past and the present, they're not future facts. They give hints of the future but they don't represent it in itself.
> **Trends have uncertain future trajectories.** One can never rely on things turning out exactly as they seem.

Trends are placed in a context and they become strategies, products or services for a happier customer and (hopefully) a better world.

As you can see, I'm keeping this chapter short. There are two reasons why. This chapter represents a kind of wrap-up of the things we discovered in the first part of the book and we are already familiar with all the terms. As we travel further there will be less and less explanations needed and hopefully, fewer question marks in our heads. Zurich anyone?

research companies
trend scouts
microtrends
macrotrends
megatrends
a trend consultant
takes these pieces
of information
intuitive approach
strategic analysis
connecting the dots
microtrends
macrotrends
megatrends
in a specific
marketing context
creating a new
innovative idea
together with the client

FUTURE ILLUSTRATION

Through presumptive activities, talk and images, the future becomes something that people can work on, talk about and imagine. Futures are not only imagined but ultimately believed in and acted on. / Jon Guice in *Designing the Future: The Culture of New Trends in Science and Technology*

Back to Hamburg, an early morning with a foggy view of the harbor. The TrendONE Hamburg office is located right on the shore. Nils enters the office and says: "Delia, Nick is going to Zurich, you want to join him?" As I have said before, I am not at all an early morning person; but it took 0.3 seconds for me to declare "yes, when, how, ja!" I wanted to meet Adrian Müller and talk to him about his intriguing research project related to future illustration. This topic was very new to me and I found it unexpectedly relevant.

Future illustration refers to visual elements that help you picture and imagine the future. Since semiotics was my favorite subject back in university, I knew how important images are when it comes to communication. Everything matters; lines, shapes, colors and people's expressions are all parts of the system that develops a story in our mind. Playing with people's projections has never been an easy task and showing them pictures of the future doesn't make things any less complicated. Back when we were children, this was more humorous than serious. I am sure you remember the famous Hanna-Barbera cartoon called The Jetsons. My favorite! It's curious to see how past generations imagined the future and how their predictions didn't come true. Where's my jet plane? Where's my robot-maid and why don't I work only three hours a day, three days a week?

It is clear that we are living in a visually driven culture. For example, the association of color with various causes was certainly a huge boom. Take for example the yellow rubber bracelets that supported Armstrong's fight against cancer. Produced by Nike in 2004 (it seems just like yesterday), 50 million Livestrong yellow bracelets have been sold. Yellow meant you are a 'Livestronger'. These rubber bracelets now come in a rainbow of colors letting people know you support one cause or another. The idea was easily picked up. "Visual dominance will have consequences – some direct, like the accelerated march of art into the consumer world; some unintended, like the decline in reading", says John Naisbitt in his book – *Mind Set!*

Data visualization is essential to information management. Companies choose to portray things, and to color them in order to reveal the data. Think for a second, how would you explain the future of media to somebody? There may be a sequence going on in your mind: first it was the cinema screen, then the TV, then the computer and now the phone. That's easy enough to understand, but if we add more details, it gets complicated. The media evolution is about going from radio and emails to web 2.0 – and social commerce to 3Dweb – and gesture control to augmented reality – and cloud computing to brainwave control – and voiceless communication. Ok, now you see what I mean? But if we visualize this…

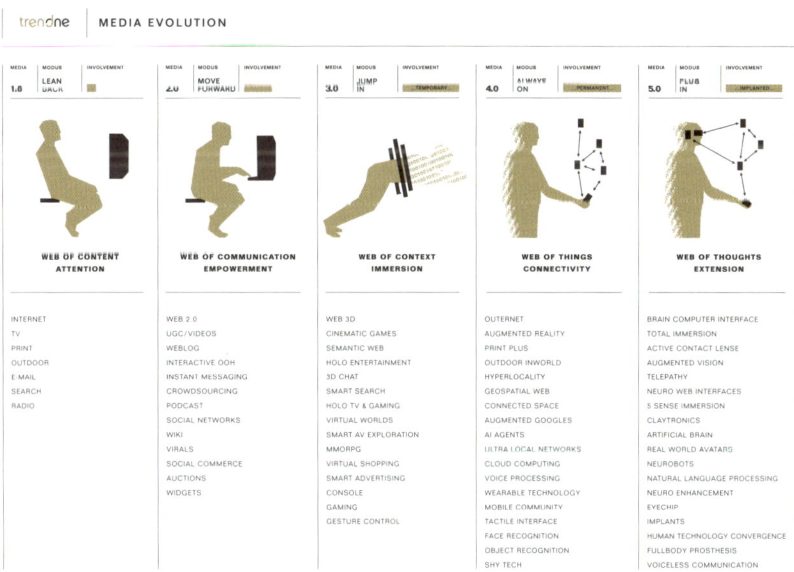

Media Evolution
/ TrendONE, 2010

…it suddenly becomes clear. The meaning of each step is comprehensible via the icons alone: lean back, move forward etc. Therefore, illustrations are indispensible when we want to give meaning to complex information. Moreover, when we deal with elements of the future, that are not here yet, the role of visualization becomes magnified. Let's face it. **It is hard to imagine something that you have never seen before, and that's where future illustration comes into the game. It's about *picturing* innovation.**

Future information (understood as information about how certain things will look like in the future) is complex. It is difficult to understand its meaning only through words; images can give the text a meaningful shape. They are faster to comprehend and nice to look at. Everybody likes to have representations: journalists, clients, your Facebook friends, everybody. Even if you don't have complicated things to say, visualization can spice up your life, attitude, mood and of course, your presentations.

Here is one example from a favorite website: Information Is Beautiful by David McCandless who recently published a book with the same title.

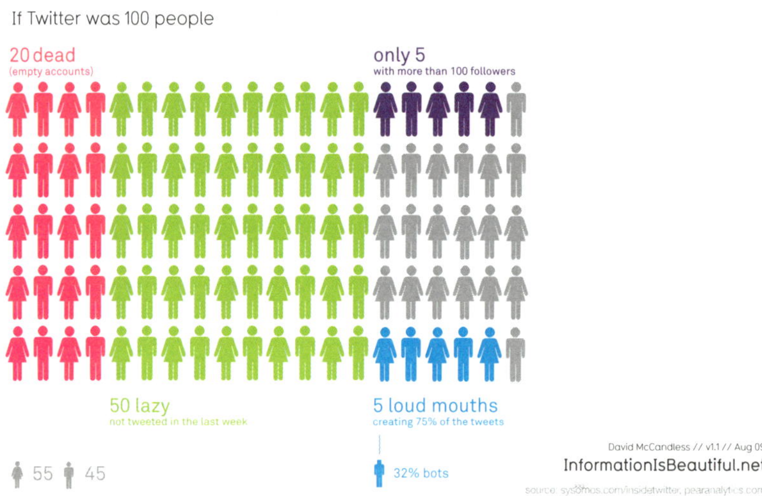

Moving apart from this idyllic land of illustration and entering serious business, there haven't been many articles written on how companies choose to visually showcase the future but it is definitely something worth keeping an eye on. Exactly the day before my flight to Zurich, the pretty Eyjafjallajokull from Iceland decided to cough up. No flight. I checked the train options and decided to take the overnight train to Zurich. Cool experience, as these trains are so comfortable and hospitable: I was even asked before going to bed whether I wanted coffee or tea for breakfast. However, the honeymoon was over when I arrived in Zurich's main train station. I had to go through five (not very friendly) people before finding someone who was kind enough to point me in the right direction. Luckily I was too excited about my meeting to get annoyed.

// VISUAL FUTURES

VISUALIZATION AS A STRATEGIC INSTRUMENT

I finally arrived at my destination and met with Adrian Müller, owner of Noocleus, a consulting firm focusing on strategic innovation. Müller is also the coordinator of the Trends Master at the University of Arts and Design in Zürich, which teaches design students trends and future research and how to use these tools in strategic innovation projects.

"In my work, strategic innovation stands between strategy and basic innovation work like design. What I realized is that the classical work moves towards innovation. We are living an era of organic growth and expansion. This means that companies have to have more knowledge about the future, their directions and innovation potential. I also noticed that a lot of design work becomes more strategic", says Müller. Take industrial design for example: if the designer creates a car he has to take into consideration values, aesthetics and functional aspects. The designer must also answer more questions related to the future of mobility than anything else. So, basic innovation work becomes more strategically oriented. This is the basic assumption behind Müller's work and Noocleus.

We had a snack in the cafeteria of the university. I love cafeterias. They bring back good memories of student life – all that freedom and randomness. But we were having a serious talk. My brain was ping-ponging from foresight to strategic methods. More about this later on – now it's all about future illustration.

Müller and I are looking at 24 in-depth interviews with leading trends and future researchers, designers and/or innovation strategists in Europe that were conducted for the *Visual Futures* study at the Institute of Design and Technology at the Zurich University of the Arts (ZHdK). The study focuses on the use of visualization in strategic trend and future research.

Adrian Müller

The main focus of the study was not on product designs, rather on visualizations of future life and innovation contexts in which most of the abstract notions come in the shape of trends, trend landscapes, scenarios, projections and visions. The questions that began to form in my mind were about types of images and their role in different contexts. What types of visual representations currently exist? In which contexts are visualizations used as leitmotifs (dominating themes) for trend and future research? Which is the semantic content of the visualizations? These are all questions that *Visual Futures* poses, so I was on the right track.

The analysis of the pictures used in the study showed that the vast majority of the participants have a very limited repertoire of visualizations, both in terms of style as well as visual medium itself. But on the other hand, everybody thinks images are necessary. Oddly enough, in verbal presentations, images still have a secondary role. They are mostly used during keynote speeches and workshops as an element of entertainment and for capturing the attention of the audience. **Most often when visualizations are used, they are duplicates of already verbalized content and don't offer extra information. In workshops, visualizations are used mainly in the creative phase: most of the time, images serve as an instrument for imagining ideas and stimulating the thought processes. When part of studies and reports, visualizations have an explanatory, almost mediating role.**

The best thing about using visuals is that they offer a wider audience easy access to new knowledge and thus allow the viewers to come closer to a complex issue. Like in the case of the *Media Evolution* illustration we had before. Even if the viewer doesn't understand all the terms, he or she can feel the meaning just by looking at the illustrations. Of course there is also the key factor of time. Who would prefer to read 20 pages instead of looking at a nice, information map that explains everything in one shot? And last but not least, the aura images can evoke. One of the interviewees from the study proclaimed: "pictures make you inspiring; diagrams make you credible". Combining them is the key.

There are various ways in which visualization can be used. Often, the images become stereotypes and/or metaphors like a picture of a businesswoman for 'new emancipation' or a 'crystal ball' for forecasts. However, during the study it was noted that a certain heterogeneity and randomness, or sometimes even a complete incomprehensibility of visualization occurred.

When I asked Müller for three of the most essential conclusions of the study he picked the following:

> Visualizations often evoke way more motivating emotions than abstract words, so they often are in an advantage concerning the stimulation and development of future related knowledge, ideas and visions.
> The use of visualizations in trend and futures research is more complex than estimated and can therefore easily be counterproductive. If illustrations are too specific (in an early innovation phase) they can destroy imaginative and creative power. Visuals of any kind could offer much more information than the persons need in order to be stimulated and the effect is reversed: the mind is shut down.
> Visualizations are best chosen if they follow a set of criteria: content (what?), context (where/how?), recipient (who?) and function (why?).

Pictures work as stimulators. They are either an aesthetic component or part of a corporate identity and have the task of creating a certain atmosphere, inspiring, surprising and arousing the interest of the viewer. Sometimes, pictures are just knowledge transferors: facts are reduced to the essential, understandable and memorable, taking the shape of an image. Other times, they can generate knowledge and increase the level of information. This is the perfect state. Patterns and new relationships between the elements are discovered and interpreted. Visualizations then become creative and analytical instruments.
The interesting moment comes when you compare the techniques to the context in which the visualization is used. For example: when should you work with photography or use a hand-drawn illustration? Is the visual constructive or disruptive? The answer to this depends on the level of the strategic process in which you find yourself.

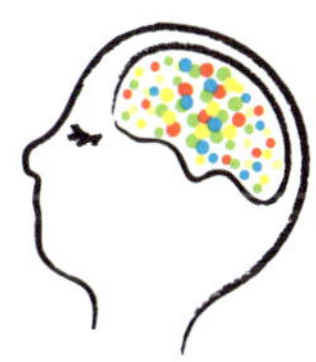

Output of the *Visual Futures* study

There is definitely a gap between the acknowledged importance of visualization and the actual willingness to use it. All the *Visual Futures* study interviewees consider visualization as being an essential instrument for strategic work. Just a few though, focus on it as a method in itself. Future illustration is a 'blind spot' in the trend and future research industry but the method has high targets and can receive professional credibility.

This ZHdK study opens a conversation on future visualization. It would be great if people focused their attention on the exploration of new forms of representation and techniques regarding different types of images. Maps in the trend and future industry should be more exploited. I am addicted to mind maps because they help you understand notions faster as the connections are revealed visually. What more can you wish for than an actual arrow that points out the connection between one concept and the other? If you are the one who builds them, the maps transform into an efficient learning tool: the visualization becomes your canvas for ideas and information and soon the connections between them will come to the surface.

TEXT ~~VS.~~ AND IMAGE

Visual and verbal language cannot play against each other; they transport information in different ways and have their own specific advantages. They don't compete: they are a team. Moreover, while language is structured sequentially, images are captured at a glance and generate immediate effect; they are cultural, transnational and understandable. While language is based solely on convention and must be learned, images not only break language barriers but also work as interdisciplinary enablers. An example of this is the Lookji app, which takes the most commonly used terms in the world and translates them into pictures so they can replace text in a message. This becomes an 'icons' message. Everybody will understand you; no matter what language they speak.

// THE FUTURES WINDOW

The Futures Window was developed in order to use the power of images here appearing as visual weak signals, to trigger employees' future thinking. It can help people work with product or market development in creating new ideas.

Rewind but don't erase! We now have to take a step back in time to Helsinki, to Elina Hiltunen. She shared with me one more thing that I purposely kept secret until now: a future visualization tool called Futures Window.

Finland already has a few organizations that visualize weak signals. TrendWiki, the weak signals database, is linked with the Futures Window allowing pictures of the signals to be visually projected on screens in an office space: lobby, cafeteria, elevator, toilet, computer screen saver etc. Visual weak signals can be photographs, animations or videos of new inventions or disruptive things that are happening today. A trend scout using this tool could also search for innovations, illustrate them and then project them as inspirations.

The Futures Window project was piloted at VTT Technical Research Center in Finland, in 2006-2007, and the results of this testing phase encouraged further development of the method. The images were projected in the lobby during the breaks of a futures workshop. Some of the reasons viewers gave for the images sticking in their minds were: "it irritated me", "amazed me", "made me laugh", and "raised positive or negative feelings in me". And it was later confirmed that Futures Window enabled the attendees to develop new ideas on how to work with technology in the future.

"When I worked with the Futures Window within organizations, I tried to show images that were related to something else than their core business, which they are already familiar with. I always tried to show images that were new and sometimes even shocking to them. I think that people have been really good in translating the signals to their specific context", says Hiltunen.

Visual images like photos, animations, illustrations, excluding normal graphs or pie charts, are used as a technique for anticipating or creating the future. A typical way to present scenarios is in a written form, but Philips for example, pioneered in using images for communicating future possibilities. In 1995, within the *Vision of the Future* project, images with possible products of the future were posted online and published in a book. Now, a video with up-dated possible products of the future that reflect social and cultural trends can be seen on the Philips website.

Emotion Containers,
Vision of the Future,
Philips Design 1996

The Emotion Containers are small personalized multimedia products, designed to be given as special presents that are meant to last and be cherished by the recipient. They have a small screen, a loudspeaker and a scent compartment. The screen can show still images or a video clip, with sound. The containers offer a more sensory way of giving, being attractive as objects in themselves and as carriers of messages of special significance.

// TRENDSKETCHING

I met Anna Luise Sulimma for the first time in Berlin at LaFutura, an event where trendspotters and innovation professionals from all over the world gather for one day to share insights. She had this kind of magical effect on everybody; people were cueing up just to get the chance to talk to her. She invented a visualization method for the trend industry called trendsketching and after this event, her career in graphic recording rocketed. It's no wonder – she is very talented.

Here is the story of how Anna created her own business, starting with her university years. While studying Fashion Design and Strategy (a course for fashion designers who want to go deeper in the field of developing concepts and marketing strategies) at ArtEZ Institute of the Arts in the Netherlands, she started a one-year trend forecasting project. It was called 'social forecasting' for which Anna conducted trend research to spot cultural and social shifts. "I collected articles, read books, did some street investigation and at the end I had loads of information. I had to present everything, my year's work, during a ten min presentation. I had to narrow down the findings to a short story, so I came up with the idea to sketch everything. That's when I drew my first map", says Anna. It was an intuitive decision in order to get rid of the confusion, summarize the information and spot the connections between single forms of information.

Anna Luise Sulimma

Anna invented a visualization method for the trend industry called trendsketching.

The development of this new method of working with trends and eliminating information overload won her the upcoming talent prize TWOTY 2009 ('Trend Watcher of The Year' award) from the trend magazine Second Sight in the Netherlands. This was the click that led to the founding of her company Trendsketcher – visual insight into complex matters through handmade sketches.

Basically, trendsketching is a visual thinking method. Visual thinking as a technique is not a new concept; it is simply about using pictures to explain your thoughts. "All human beings think in pictures. Most people's brains work visually. That's how we remember things, how we see our world – with our eyes – and save it in this way", says Anna. Therefore, she took this basic role of the image and placed it in the trends and futures industry.

Trend research is always about amassing information then narrowing it down to what is most relevant in order to see the connections between the facts, so this tool is both useful and necessary.

Talking to Anna made me feel like I'd just spotted something special. You know when you discover something and you can't wait to tell others? It was the first time I had heard about an 'instant trend graphic recorder'. When companies have workshops or special meetings, regarding their future strategy, Anna sits next to employees/clients and creates a visual protocol of what is happening. She basically illustrates the content of what they talk about during the meeting and its outputs. This way, no important thoughts are lost and they can see where they started and what the conclusions were whenever they wish. Her sketches are as simple as possible and as complex as necessary.

Anna is also an expert on creating short animated videos based on certain trends or developments. She collaborates with the Zukunftsinstitut (Future Institute) in Germany and creates future videos for the Institute's clients. These videos are then used as an inspiration for employees to develop marketing strategies based on the new information provided.

Being eager to hear more from Anna, I call her one late afternoon and I start shooting my questions like for a Q&A magazine article:

How do you choose a topic when starting a sketched trend analysis? "I first look for something that triggers me, something that irritates me and then at the status of today's topic. Then I take a step back in history and try to find out where the trigger comes from. Next step is going into the future, thinking of what could come out of this piece of information, how could this phenomenon or trend be translated into a product or a great business idea".

What are your research methods? "Online research, flipping through the press, or what I call natural future research fields – always keeping my eyes and ears open. And I always try to find books…because they help me get deeper into the topic. But it also takes more time. For example, I did some research on 'the future of men' so I read a book called *100 Years of Menswear* to get into the history but I also read other information in regards to how to become a father these days. Zeitgeisting! I also gather information from big research groups like Nielsen Research Company who publish reports on people's behavior in a certain group, getting a bit of statistics".

Are statistics relevant in the case of trend visualization? "Somehow, everything comes together. I need some numbers as a background because most people are very interested in numbers. From my point of view, the qualitative research is more important. It's more about intuition, how I think and feel about how things will develop and not so much about statistics".

Why don't you believe in statistics? "Because statistics are most of the time somehow fake as they nuance the question. Depending who you ask and how you ask, you can somehow always get the answer that you wanted. This is not futures research".

So how do you bring innovation further with your method? "It eases the process of ordering information and it works for everybody who can see. Building scenarios with visual tools makes you literally see what might happen".

What does trendsketching as a tool mean for the future? "On the one hand, I think this method will spread more in the shape of animated images because technology is also developing. When we are able to use 3D visualization then it will be even more efficient. Imagine if you build a trend map where you can see what is happening and how things are connected in three dimensions. It would be a more realistic model of the world. In the meantime, I think visualization is a great tool that can be used to help us take a break from technology: working only with people, pens and paper."

Anna hopes (and I do too) that companies and the leaders of companies will be braver in trying a visual method no matter if it is hers or somebody else's. Some conservative companies are taking a step back saying that "the only good word is the written word", but progressive ones see how these maps can help keep track and highlight changes that are in the planning stages. Hey, and maybe her maps aren't a classic form of decoration, but I'd surely put them up on my wall and stare at their complex beauty all day long.

// ORKIDEE. TREND TRANSFER

I somehow felt full of energy while researching the future illustration topic. I was so inspired just by listening to my interviewee's lovely stories and current activities. It was a brisk Friday afternoon right before the start of winter and I was on my way to Orkide Ergün's studio in Neukölln, Berlin. It's a neighborhood founded by bohemian religious refugees in the 1700s and still retains an old-time village feel. Currently it's most populated by Turkish people who spice it up with their cuisine and essence.
Orkide is Turkish as well. She was born in Istanbul but moved to Germany when she was a child. Her company ORKIDEE – 'an orchid with ideas' – specializes in trend-based branding. To me she has one of the most visually stimulating jobs in the world, so I was keen to find out her story. We started with, of course, making Turkish coffee in her studio's kitchen and moved on to one of the rooms where she is usually in the process of reading, tearing-up magazines and then creating mood boards.

Orkide Ergün

TREND ROOMS

Orkide studied product design within the textile design department at the University of Arts and Design in Hanover. The first time it dawned on her to use trends in her job was when she crossed the path of Li Edelkoort – a trend forecaster, curator, art director and publisher (to name a few) based in Paris. She is for Orkide 'the queen of trend research'. Not only for her, Edelkoort is renowned worldwide as a soothsayer in the world of trends; what she predicts is taken as concrete and applicable, no questions asked. This is all due partly to her seemingly innate talent for visualizing trends. Edelkoort works with everybody, from the most cutting edge technology firms to 3rd world artisans, using them as sources of inspiration and assisting them as clients. Just check out her website and I'm sure you understand this following quote from Orkide.: "I was so inspired and impressed by her. It was in 1996 at the Frankfurt Textile Fair for Furniture and Interiors when I first saw the so-called 'trend rooms' that created 'trend worlds' made by Edelkoort's team". Soon after this, Orkide began working in the design team of the Frankfurt fair for three years.

The term 'trend room' sounded inspiring. I asked for more information. 'Modern green', 'classic yellow', 'brocantery blue', 'exotic red' and 'neutral shock' where themes given to Orkide in order to create a room using all the fabrics from the fair. She had to paint and express trends with colors and materials: tapestries, wallpapers, curtains, and cushions were her brushes. After moving to Berlin, Orkide worked in a team at dan pearlman, a strategic creative agency in Berlin, with the BMW Brand Academy where she created rooms for each car in order to help the sales people understand the product better. "The sales people had to become part of the brand; they had to feel it with all their senses. [How does MINI smell!?] We didn't only have to think about what the car looks, feels and smells like but we also had to combine this with information about the MINI clients: where do they live, what do they listen to, where do they eat. We got a brief and we made it *visible*", says Orkide.

MOOD BOARDS

ORKIDEE is not doing mood rooms anymore. She switched to mood boards, visual trend concepts used as an inspiration for branding. I was wondering what led her to this decision. "Li Edelkoort is very abstract, artsy, sometimes you understand her, sometimes not. Her language is understandable for designers but for people who are not involved at all in the creative phenomena it is a bit more difficult. So I decided to make everything more concrete, showing the applications of trends", says Orkide. She soon started to fill this gap between the artistic and the practical world. During her studies, she created a trend book called *Skin Knowledge* where skin (hence the title) was the trend. Remember how popular it was, in the late 1990s, to get pierced, tattooed and/or show a lot of skin? Botox and aesthetic surgeries soon became popular as well; therefore skin was popular. "As skin is feeling, I said to myself that I will do a feel book with mood boards because during the day we unfortunately only feel the keyboard of our computers and we don't do anything with our hands anymore", explains Orkide (by the way, that's why I started knitting). The *Skin Knowledge* book activated the tactile sense and served as an inspiration book for the textile

Mood boards from the *Skin Knowledge* study / Orkide Ergün

industry, fashion, product and interior design, architecture, cosmetics. It looks like an exquisite schoolbook. Paper, organza, colors inspired by the image of skin under the microscope, transparent materials, and other textiles were all made by Orkide's hand. For the first time in an easy to follow format, she was translating trends into visual inspirations, visual concepts. And this is what she is doing for her clients today as well, just in a more concrete and articulated way.

When you flip the page you'll see one of Orkide's mood boards from the 2005 furniture fair in Germany (not allowed to say which one). Its theme is 'chalk color' and it represents a color palette as an inspiration for anything furniture related, including textile. Why non-colors? Back in 2005 there was a social uncertainty floating in the air so this is reflected using undefined colors. You can't tell if one is whitish or grayish, they are ambiguous hues and tones. The inspiration was Rachel Whiteread, a British artist, best known for her cast-like sculptures. Stone, sand, raw materials are all elements of this mood board. It has a natural and calming look don't you think?

Chalk Color mood board / Orkide Ergün

WHERE TO BEGIN? THE CLIENT AND THE BRIEF

"Sometimes I just get a brief: next fragrance will be summer fun for 2012 and the description of the target group. So then we start thinking what summer fun means, how does this target live in the summer, is it more about fun or joy?", says Orkide. That is quite…abstract, right? The client might not even have the name of the product before briefing Orkide, only the idea that they want to do 'something with summer fun' for example. The mood board becomes an inspiration for every aspect of the product: the bottle, the advertising and marketing concepts and sometimes even for the fragrance itself. The client then receives four different trends from Orkide, in the shape of mood boards that explain and play with the meaning of summer fun. The trends are totally different from each other; one of them is chosen and recommended for further use. Everything starts with the boards. Later on, these elements are visible in the overall look of the product and its advertising campaigns.

This is when I first find out that perfumers follow trends as well and that there are trends in fragrances too.

TRENDS IN MOOD BOARDS

"Microtrends and consumer trends are to be found in mood boards, as it is very important that clients see how the topic of the board is expressed in today's society, in movies, books, art. I find the information in magazines, on the Internet and by reading blogs", says Orkide. The piles of magazines lying on the shelves in the back of the room reflect her consumption of all the cool media in existence. She adds, "Art, exhibitions, shows, fashion shows, interior fairs, design fairs and every day events. Sometimes it is very funny how I never shut down, I am doing my job all the time and I am always analyzing any picture that I see on the street or any talk show that I hear on TV. Another important part of the research is the store check. Window shops are trend indicators and especially in Berlin as there are a lot of flagship stores".

It's important to hunt for new images all the time. If you compile key pictures of the moods you need nothing else to describe a phenomena or trend. It gets decoded, like in semiotics. Orkide enthuses, "Everything is clear when you look at mood boards, long texts are not needed! On the other hand, no concrete ideas are told because their role is to be an inspiration and the images should be interpreted in various ways".

Consulting with mood boards is interpretative work and before starting to actually build them, you have to know what trend is behind the concept, its core value. The mood board is a knowledge transfer from the trend to the concept. Orkide goes on to explain, "Creating mood boards is like painting because you need the key frame and then you have to put the elements [the pictures] inside in order to create a coherent and inspiring composition. Looking for pictures can take days and days. This is the most difficult part, as the power of pictures is amazing, you never forget pictures. You can't communicate only in words, if you see images, you can't ignore them. The power of the pictures is heavy together with keywords".

TREND TRANSFER

Together with hello-hello, a non-traditional marketing firm that translates visual and cultural trends into working concepts and strategies for brands, owned by Judith Mair, ORKIDEE does 'trend transfers.' Their latest project comes in the shape of a book called *How do Megatrends Influence the Cosmetic Industry?* It explains the influence and impact of four megatrends on products, design, marketing, communication and sales within the cosmetic industry. The study provides insights, information and inspiration with the aim to bring together future innovation and strategy for brands and markets to become visible.

I MEGATREND: HEALTH – HEALTH HYPE
II MEGATREND: SUSTAINABILITY – GREEN REVOLUTION
III MEGATREND: AGING – DOWN AGING
IV MEGATREND: PERSONALIZATION – SOCIAL INDIVIDUATION

Take a look at the mood board on the right. It represents a design inspiration for the megatrend 'Sustainability.' The trend transfer has been made from the economical, cultural and social behavior to a visual board, a mood board that serves as an inspiration for the cosmetic industry. The colors, textures and forms can be incorporated into the packaging, container and/or the attributes (the color or consistency of a cream for example) of a cosmetics brand. "Green changed from electric to yellow green, eco developed into sustainable and so it became more refined and smooth", says Orkide.
When I think of fashion and beauty I see them as fast moving industries that are inspired by microtrends, which they pick up and immediately deliver to the masses. However, this study offers general megatrend information and targets group descriptions with inspirational mood boards. Orkide talks about megatrends as something that should be a fast moving market. Why? "Megatrends are very good for new ideas but if you create something very specific [a new product for example], then you need to go deeper into things and you need additional research and analysis [spotting microtrends and macrotrends as well]. The study shows how we work and what we can do.

when we work for a special brand, then we have to do the research process again and again and that can't be sold to anybody else because it is very specific", says Orkide. So, when it comes to fast moving markets, ORKIDEE doesn't consult based on megatrends but starts top-down, tailoring the megatrends on specific situations in order to find their relevance by looking for the roots of the trend.
"In the future, nothing will change. Not my technique because the technique itself is the image and it will be forever powerful. Maybe the networking will be something that could change, innovative collaborations could arouse. I think I'm going to do mood boards for years and years", exclaims Orkide with a big, beautiful smile on her face.

It is time for me to go now, though Orkide's studio is so inspiring and cosy that it makes me wish I could stay a while and flip through some magazines. But I need to go and get ready for my trip tomorrow. Also, regardless of the late hours, since I am a so-called 'trend researcher', I need to fake that I still have a social life and stop for a beer with my friends.

‚Sustainability' mood board from *How do Megatrends Influence the Cosmetic Industry?* / ORKIDEE, Hello-Hello

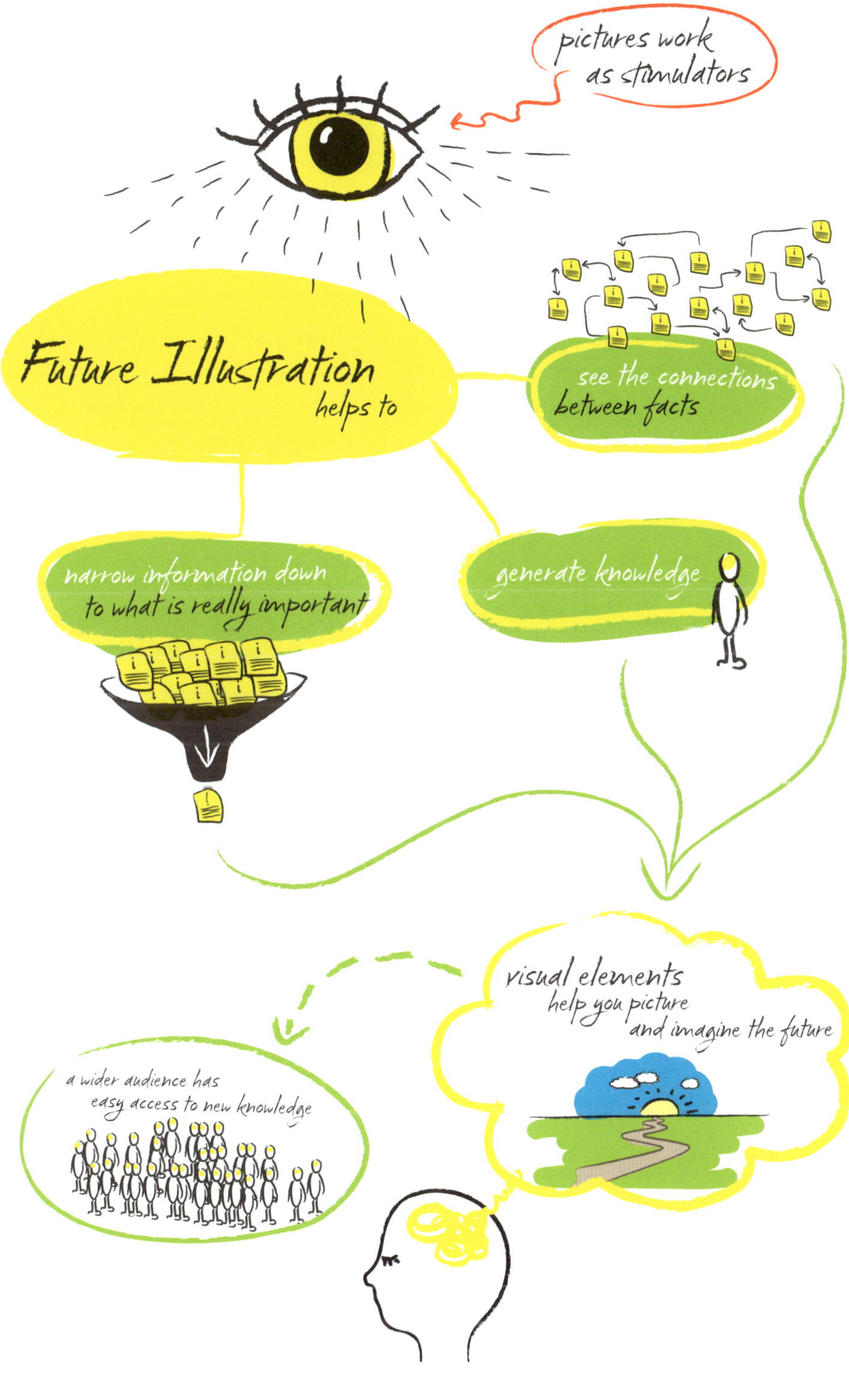
pictures work as stimulators
Future Illustration helps to
see the connections between facts
narrow information down to what is really important
generate knowledge
visual elements help you picture and imagine the future
a wider audience has easy access to new knowledge

THE INNOVATION COCKPIT

Rolf Buchholz

We all have prejudices. Or better said, we all use stereotypes. Especially regarding certain countries, their people, culture and customs. These stereotypes can be positive or negative and/or accurate or inaccurate, but they all make assumptions. Here ~~is~~ was one of mine: German trains are always on time. So, here I am traveling to Wedel, a small town near Hamburg, trapped on a train that's running late, for my meeting with Rolf Buchholz, founder of KEY VALUES and lecturer of Innovation Management at the Hamburg School of Business Administration (HSBA). I like trains, so I am secretly enjoying this delay.

Where were we? Ah! Yes. We started talking about placing trends in contexts. **We have now learned about trends and consumer insights but what do we do with this information? How do we choose what's relevant, how do we deal with the mass of information, what type of criteria do we use to chose this information?** Things need to be evaluated and put in a context. One way of managing this heap of information is using IT supported data management tools and Rolf Buchholz came up with an interesting new device that fits the description: the Innovation Cockpit®.

The name sounds somehow supernatural, doesn't it? Is it a bird? Is it a plane? Is it a time-travel machine into the future? If you guessed 'None of the above', you are absolutely correct! Fancy as Innovation Cockpit® sounds, it is not a machine that can swallow information and give solutions on the spot, but a human process instead, of course supported by IT.

Let's start from the beginning. When creating KEY VALUES, Buchholz knew that innovation was fundamental to the success of every corporation. We also know that. What's new for us is this statistic: the rate of failure, in areas like FMCG (Fast Moving Consumer Goods like food and cosmetics), is beyond 80% in terms of new products and services introduced on the market. Spotted: huge gap between costs, time and leadership. Buchholz's idea was to close the gap and leverage the space for improvement. Therefore, "KEY VALUES is positioned in the service industry, focusing on innovation and business development. We consider ourselves as a value creator and a value driver. Our aim is to help our customers to better reach their objectives and empower their competencies" says Buchholz.

KEY VALUES cooperates with trend and research companies and when it comes to the execution of their proposed solutions, they work with research & development departments, IT departments, as well as with design agencies. Their working model focuses on the integration of tailored methodologies, insights, competence and creativity in order to give innovative solutions. Based on those, KEY VALUES developed a series of tools and frameworks that are individually applied depending on the company, situation, objectives and culture. All of them have different aims but today we will concentrate on the Innovation Cockpit®.

Buchholz explains, "The Innovation Cockpit® represents a data based controlling instrument for innovation development. We found out that decision makers have too much information to cope with at the same time, so they might not see the missing critical information that is actually relevant to them. And they definitely do not have the time, sometimes neither the know-how, to narrow down the bulk of information to a target oriented essence".

The context of a company always changes: business models, channels, behaviors, attitudes, technologies, usages, lifestyles – you name it. Companies need to deal with dynamic and fast changing environments in smart ways. A professional could help a company understand the contexts they are working with and develop concepts that imply the current market dynamic and that's when the Innovation Cockpit® comes into the game.

THE CUPCAKE CASE

Buchholz tries to help me understand the Innovation Cockpit® thought process by giving me the cupcake example (I'm paraphrasing here): A German baker comes to us because they are having severe problems getting additional listings from their normal distribution centers (stores). We notice, in London, up-scale cupcakes are all the rage and people have no trouble spending a lot of dough on them (pun intended). Seemingly unrelated, the luxury segment in Germany is still growing despite the financial crisis. Even so, let's say that there isn't any luxury bakery product in Germany. In this situation, we can clearly build the following conclusion: we should create a designer bakery in Germany.

You would probably say, "Ok but I could come up with this idea on my own, why should I go through a long innovation process instead?" Setting up a designer bakery brand is an idea which can only become relevant (aka cutting risks of failure) when it is driven out of individual relevant information settings: how the bakery scene develops, trading development, trends and growing market segments to name a few. So, the Innovation Cockpit® is not only about being creative but is also data driven. Opening a luxury bakery in Germany could be a complete failure due to the type of culture. Maybe German people are not so fond of fancy things and prefer more practical ones. One needs to know everything before venturing forth!

Now let's say in Germany cupcakes are all of a sudden a snack people can't get enough of. They find them tasty, hip and cool and are willing to pay a pretty penny for them. Through the Innovation Cockpit® process KEY VALUES could generate the idea for the German baker to enter a different distribution channel like the confectioner's shop. Then, they would not only be present in the consumable goods retail trade but also in the high-end café and bakery trade. Their clients would perceive their product as a delicacy, thus new target groups would be attracted.

Normally, the research and implementation of innovation is much more complex than the aforementioned example. In real Innovation Cockpit® example, KEY VALUES would combine various forms of information and spot the impact of specific information: positive, negative, huge potential, risk, idea generator, insights. Starting from there, they would develop strategies and concepts. At the end of the Innovation Cockpit® process the consumer leaves with potential innovation routes that are basically small innovation concepts to be tested in the consumer field. An example of an innovation route here would be to create a new brand for a luxury bakery company by using the same pattern one finds in the fashion industry: create bakery collections that change over time, that change depending on specific lifestyles, themes, topics and so on. What leads to the combination of information is easy to notice here. The Innovation Cockpit® process didn't only identify the luxury segment as an innovation route but also took ideas from the fashion industry to combine it into something homogeneous that can be tested and has a chance to breakthrough.

As you probably already noticed, my favorite reoccurring theme on this road trip is when we discover how trends, in any shape or form, are related to or involved in the innovation processes. Sure, trends are only the start but they are constantly an inspiration and I wanted to know where and how would I find them in the Innovation Cockpit®. I paused for a while and asked in a very serious tone, "What kind of trends does KEY VALUES deal with at the research phase in the Innovation Cockpit® process?" The answer: with all sorts of trends, but mainly with market dynamics (macrotrends) such as consumer insights, values, upcoming laws, technologies, patents etc. They are all constantly fluctuating therefore it's essential to take them into consideration. If you miss any market changes, you are then dealing with a great risk to your new product. If you don't take market dynamics into consideration then your cupcakes may end up going stale on the shelf.

// INSIDE THE COCKPIT

Then, I was wondering, what kind of prerequisites does a company need in order to begin the Innovation Cockpit® process? Buchholz says, "Initially, it's essential to understand the culture of the company to decide upon their way of innovation". The road to innovation, my friends, has to be a chosen one. Buchholz adds that further key aspects are, "understanding the driving forces for innovation and considering changing factors that influence the business, products and services".
Let's take this process in baby steps.

The check-up or better said the data evaluation and environmental analysis. The Innovation Cockpit® starts with a brief analysis or a status quo set up, looking at what kind of information the customer receives on a regular basis. Reports, newsletters, research information. "It doesn't make sense to enter the Innovation Cockpit® if you don't have any idea about where the world is going. This is prerequisite no 1: the right information has to be available in the right kind of format", says Buchholz. He also mentions that, sometimes companies buy information that is necessary for entering the Innovation Cockpit® process (trend reports, consumer insights, trade information) but have not reviewed it in years or they may even forget that they currently pay for and receive this kind of service. Unbelievable!

Narrowing down the information and setting up the specific tasks. Obviously, there is not enough time for evaluating the whole world. There needs to be specific objectives. For instance, this step could include the evaluation and innovation potential for a specific product group (shampoos) or market group (shampoos for young people) or given specific circumstances (shampoos during winter time). This objective is decided during a discussion with the people responsible from the client company. Talking is key, always and forever! "We have an initial talk with the company and we raise critical questions: if they say we need something new we ask what kind of new. If they say they want to make a line extension we say ok, you can do that without the Innovation Cockpit® [as line extensions means adding services or products that correspond to already existing targets, values etc.] If they want to develop real innovation then we ask in which direction: is it for the product, service or distribution channels? What do you want with your brand, what's your strategy?" says Buchholz. Obviously, you can only reach certain goals if you have an idea of what you are looking for. This reminds me of a very funny Romanian-style joke my father told me once: Bula was praying every night to God to make him win the lottery. One night, God appears in front of Bula's eyes and says, "Bula give me a chance, buy yourself a ticket!" It's the same with innovation. Get yourself an objective! A relevant one! The discussion with the client leads to a single objective, a target line that the customer has to approve. "We are sometimes in the situation when it is not so clear what the customer really wants. At the end of the day, we have to have a clear target, no matter what!" exclaims Buchholz.

Detecting the innovation drivers. Ok, now that we have the target, what do we do with it? This is where the hard part comes in. We need to detect the drivers and the influencing factors of our objective. What does KEY VALUES mean by drivers of innovation? "Innovation has aspects that accelerate the need of innovation. Meaning, consumers have a certain issue and innovation helps them to overcome it: a new technology that has been developed or a certain application", says Buchholz. This is precisely why the information offered by companies such as TrendONE is so valuable; they are actually selling drivers of innovation.

Out there on the market there is an incalculable amount of innovation drivers and inhibitors, which renders it impossible to monitor the extent to which they influence the innovation management process. And this is a human process too. When you have the results of your research it is time to determine what is relevant regarding the objective. This means extracting the important information and placing it in your specific context in order to discover common fields or differences. At KEY VALUES, this is an analytic internal process. "We condense the target and filter relevant information that we place in a certain context. This will lead us to a potential that wasn't clear before; it will lead us to new ideas and generate new insights", says Buchholz. This analysis then becomes scenarios, innovation roads and planning of how to apply the specific strategy, the new idea.

After this, there comes the testing. "Everything ends with consumer testing! It is very relevant", says Buchholz. This is interesting! As you remember we also heard that everything starts with consumer insights in the chapter with the same name. I can see what he means: there is no clear accepted theory regarding the innovation management process. In most cases part of this process is called the Stage-Gate process. Pioneered and developed by Dr. Robert G. Cooper, the Stage-Gate model is the world's most widely implemented and trusted product innovation process. According to its website, "Stage-Gate divides the effort into distinct stages separated by management decision gates. Cross-functional teams must successfully complete a prescribed set of related cross-functional tasks in each stage prior to obtaining management approval to proceed to the next stage of product development". Wikipedia sums it up: "These production areas between the gates are idea generation, establishment of feasibility, development of capability, testing and validation and product launch. At the end of each of these steps for developing a new product the senior management makes a decision as to whether or not the product should continue to be developed".

You enter the Stage-Gate process with the collection of relevant ideas, influencing factors and trends. Then the validation phase follows. **Validation means checking whether this preliminary information fits your brand, distribution channels, credibility and potential.** From here you develop the innovation routes or concepts that have to be tested. At the same time, a feasibility test with the R&D department needs to be done – deciding whether the idea you have in mind is realistic and can actually be produced. However, according to Buchholz "the Stage-Gate process is rather conventional wisdom, which doesn't reflect the 'real-time dynamic' but a sequential method. The modern innovation management processes consider that the latest information should be weaved in on a constant basis". Most companies have created their own innovation process that is tailored to their organizational structure; and it is never the same. The Innovation Cockpit® is one tool and framework within the innovation management process but it is not self-sufficient. Building innovation can't happen exclusively by using the Innovation Cockpit®. It is a device that ensures professional innovation management and effective innovation development.

MORE INTERESTING KEY VALUES TOOLS

I was also interested in other tools that KEY VALUES uses for building innovation in companies. I like the sound of one of them called the Business Innovation Scan, so, even if I was running out of time for this meeting, I took a turn from the main subject and asked for its details.

There is always the need for relevant information and insight to be discovered. The Business Innovation Scan is a process that addresses midsized companies. They are sometimes very innovative but don't have the proper methods needed to grow. In fact, most of the time, they don't have innovation departments. They probably have a few marketeers that do everything from business cards to outdoor ads and websites. With the Business Innovation Scan, KEY VALUES goes inside the company for a few days, looks at the potential areas for innovation within the company, looks at the processes, co-operations, partners, brands, products, services in order to identify new opportunities. Then builds a report for the company's management in order to show them where they could innovate – the innovation routes. What they get from this scan is a complete overview of where the low hanging fruits are in terms of innovation and what they could do in order to grow. It helps midsized companies identify their innovation potential regardless if it's driven from their competencies, their brands, service opportunities or process enhancements.

Business Innovation Hub and Brand Cockpit are other KEY VALUES tools that I am intrigued by, but let's save them for the next book.

// THE FUTURE OF THE INNOVATION COCKPIT

As we probably all imagined, it can be difficult to involve clients in such projects. I was wondering what kind of arguments one has to make to convince clients to do something really different with their company. How do you persuade them be 'courageous'? Confidently, Buchholz answers: "We don't convince clients. They convince themselves... Our experience says that you are capable of innovation if you are capable of change. One depends on the other, that's simply a consequence. A client cannot expect major improvements or innovation that generates revenue streams without taking calculated risk, leaving the comfort zone, thinking and acting different than the others". It is KEY VALUES' job to help customers through this difficult process and influence everything they can to convert calculated risks into success.

I was, of course, curious as to how Buchholz sees the future of the Innovation Cockpit®: "In 5 years from now, marketeers will go into the intranet for the Innovation Cockpit®. There, they will find the visualization of an airplane cockpit. All the information that is relevant to their brand or innovation process is outlined there. Which are the latest trends in my product group? Which are the successful innovations and consumer expectations? This will be an automatic process. Imagine you have different instruments appearing on a simulated airplane control panel and you can look at them to find out the so-called critical aspects: consumer's attraction regarding bio products (decreasing or increasing), categories in the trade (products are listed out, others coming in), and product categories popping-up etc. Buchholz adds that "various research and trend companies will deliver online information that feed a client who will simply stay and drive the innovation process on his own, in his cockpit." As imagined, the significant things will flash as in a real cockpit: Attention something's happening! The good thing about this idea is that companies will no longer ignore or forget about the information received. Everything will be visible and centralized, including information from different countries or product groups. Clients will see how their brand or product range is doing in one country as opposed to another or whether a competitor is approaching.

Innovation development is for many companies an unknown flight and one of the most important and difficult tasks.
Whatever shape it takes, innovation is essential for growth and key for companies to remain successful across all branches. In this respect innovation will always be a critical success factor, although methodologies and approaches will evolve and vary. "I believe we will be creative and create new buzzwords for the same thing. [This is already happening!] There is almost no tradeshow where you don't find the word 'INNOVATION' on the boost. As we need to be innovative, we maybe need to create a new word for it", says Buchholz…Oh please no! It wouldn't be fair to anyone; not clients, nor the readers of this book. And imagine I would have to start all over again and rewrite this text. The innovation field needs law and order. Not in terms of who does what and super-organized creative sessions, but firstly in terms of language. It is very important that we all start speaking the same language – the innovation language.

Hence, conclusions being drawn, I am taking a quick break until my next interview begins. I have decided to do this one on Skype. All the travel in the past few months has gotten to my bones and I need to spend some time at home in Berlin.

"Knowledge is the sum of many pieces of information. Current data is therefore crucial for success. Only those who know the latest market and consumer insights, trends and market dynamics, micro-analyzed, interpreted the drivers as barriers correctly can innovate effectively.

/ Rolf Buchholz"

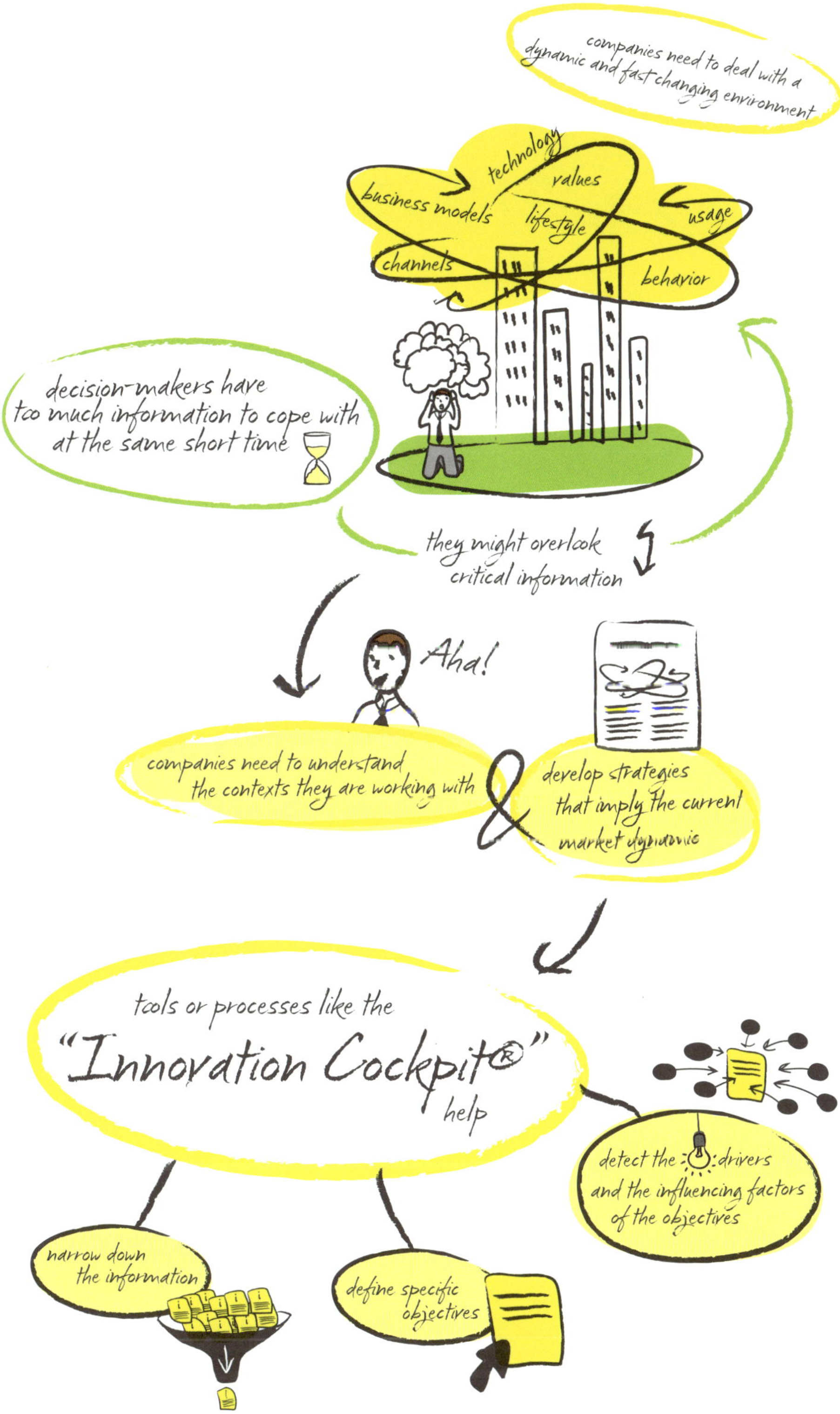

companies need to deal with a dynamic and fast changing environment
technology
values
business models
lifestyle
usage
channels
behavior
decision-makers have too much information to cope with at the same short time
they might overlook critical information
Aha!
companies need to understand the contexts they are working with
&
develop strategies that imply the current market dynamic
tools or processes like the
"Innovation Cockpit®"
help
detect the drivers and the influencing factors of the objectives
narrow down the information
define specific objectives

SCENARIOS & WILD CARDS

I am in home-office mode for the rest of the week, so I decided first to take a short walk to the Turkish market by my house, a place of oriental enchantment and culinary endeavors before I got started on the all-consuming yet enjoyable next chapter of this book. I arrived back home with my arms full of goodies, dropped them on the table, sat down and started typing away. The theme of the week is scenarios and wild cards. I slowly dive into the topic and feed my brain with the basics. There's this article I discovered by journalist Giselle Weiss in *The Global Investor issue 2.09 Focus_04* in which she asks Richard Watson, publisher of *What's Next*, the key question "Trend research only analyzes developments that are already apparent in the present. How do you get from the present to the future?" Watson's answer was as insightful as possible. He said "Imagination and curiosity. You have to constantly be asking questions: How might things intersect? Where could things go? In a way, it's very speculative. A lot of what I do is extremely silly. But the more serious end of it is scenarios. You should never believe a futurologist who says, 'The future is X.' The future is by definition uncertain because it hasn't happened yet. According to that logic, there must be multiple futures. **Scenario technique is a wonderful strategic tool because it actually acknowledges the fact that there are multiple futures**".

The scenario technique, an ideas generator and long-term look-into-the-future strategy tool, dates back as far as men have been fighting wars. While working for the RAND Corporation in the 50s, Herman Kahn – a preeminent futurist of the 20th century – brought the scenario technique previously used for analyzing the likely consequences of nuclear war and adapted it to the business world. But the most famous company identified with scenario planning is Royal Dutch, Shell's Group Planning Scenario team. In the 70s, Pierre Wack, an unconventional French oil executive, saw the need for long-term planning to anticipate changes in volatile oil and gas markets. When the first oil crisis hit home, the value of Shell's futuristic approach was realized. The oil embargo caught most companies by surprise, but Shell had already considered the impact of an increase in oil price as a possible future scenario and thought of the actions they should take if it happened. By practicing on this scenario they managed to avoid the worst shocks; Shell even emerged after the oil crisis as the strongest player in the field. What is really amazing and useful about the scenarios tool is that it allows the inclusion of factors that are difficult to understand, such as new insights about the future (e.g. wearable technologies), deep shifts in values, unprecedented regulations (laws) or inventions (e.g. the bionic eye). Somehow, the relationship between these complex factors becomes obvious. To paraphrase from Wikipedia: when disclosed years in advance, certain weaknesses can be avoided or their impacts can be reduced. A company may discover that it needs to change contractual terms to be protected against a new class of risks, or save money to purchase anticipated technologies.

I dug further and, from crowd-sourced Internet platforms, I found out that **the scenario technique works with at least two types of information: the things we believe we know something about and elements we consider uncertain. The things we know are the trends in whatever shapes they come: micro, macro or mega.** As the scenario technique is a method for looking into the long-term future, megatrends are the main type of trends used in the process. By looking at them, we can safely make assumptions about demographic shifts and technological developments. The uncertainties could be anything from the context that surrounds our topic to the outcomes of political elections, innovations, fads, fashions and so on. "The art", as Richard Watson spoke about in his interview, "lies in blending the known and the unknown into a limited number of views" (aka scenarios).

Elina Hiltunen wrote an article called *Scenarios: Process and Outcome* which was published in the Journal of Futures Studies where she states that, "scenarios help us to be prepared for alternative futures and they question our persistent beliefs about the futures". In the same article she adds, scenarios help us to "innovate the future by breaking our mental models and encourage us to create something new". A colleague of Hiltunen that worked with the scenario technique for years commented that scenarios are like "wind tunnels for strategy". Another car related metaphor could be test-driving. "The process, the phase of making the scenarios leads us to seek for information about the future, weak signals, trends, wild cards etc. and think about the possibilities of combinations of these elements", says Hiltunen.

// UNDERSTANDING HOW TO WORK WITH SCENARIOS

According to Herbst, there are several hurdles that have to be overcome in order to achieve these goals: accept that the environment changes continuously, identify the relevant developments, interpret their implications and make decisions for a new strategy.

All this research was done in order to have an intelligent conversation with Michael Herbst, Business Unit Manager at UNITY – a management consulting firm on strategy, processes, technologies and systems, who has been working with scenarios since 1995.

I have read many articles about scenarios but only after my talk with Herbst did everything seem to come together. Basically, UNITY's understanding is that **future knowledge is already a commodity**. This is interesting! We live in a knowledge society where access to information is no longer hard to get. Information is available at any time, everywhere. To compress information into knowledge is almost no effort at all. Piling up expert knowledge was considerably slower before the help of the Internet and Wikipedia – it was only possible by spending significant time gathering information in libraries. Therefore, UNITY's core business is using compiled information – the future knowledge to cause organizations to innovate and focus people's mind-set towards the future. One of UNITY's tools, which helps to achieve part of these steps, is the scenario technique.

INSIDE THE 'SCENARIO GRINDER'

As I am a visually driven person, to help my brain understand this correctly, I imagined a rather small meat grinder with similar properties to Mary Poppins' bag. So, endless possibilities with regard to what you put inside. But, what do we have here inside the 'scenario grinder'?

Trends as influencing factors. UNITY extracts the relevant aspects for their customers from the information received from other companies that spot trends and make future predictions. Remember I said this is more common than you might think? They first aggregate the available future knowledge (trends and other information they get from research companies) into several perspectives on the future (aka scenarios). The next (and more important) step is to understand what these developments and changes mean for the company. (Digital camera become people friendly and consumers prefer to use digital cameras over analog ones).

Insights. Where are the threats and opportunities for the specific business in the contexts influenced by the factors? These insights offer output orientation and aid in decision making for future innovation fields. Like Kodak taking digital photography as a change factor for their business. (Digital photography has become more and more popular).

What comes out on the other side of the meat grinder? Herbst answers, "Scenarios that might influence the future of a company we are talking about. It's about asking yourself, after grinding all the data (aka clustering and spotting the connections between factors), 'what if...this or that happens... how would our company react?' or 'if this happens how can we prepare now for these possible future changes?" (If digital photography increases in popularity how could Kodak change its products in order to stay popular or how could Kodak innovate in this context. If digital photography is popular then Kodak would do x,y,z).

GETTING INTO DETAILS

When UNITY talks about the scenario technique it means two things:

Michael Herbst

1. **Multiple futures:** we allow alternative developments for the future and don't predict only one.
2. **Network Thinking:** we consider the dependency of factors that influence our industry.

At the beginning of a project they try to understand the influencing factors (aka trends) for the specific topics. This step is taken together with the customer; its purpose is to find out what will drive the future. Normally, UNITY has a huge catalogue with influential factors: 80 to 100 and they tailor them to every situation. "Our approach is to develop the future together with the customer; we don't develop the scenarios and then show the futures to the client. Why? Because it might happen that you tell the customer things that he already knows. We believe that the scenario technique has a lot to do with creating a shared vision among the group and the company", clarifies Herbst.

80 to 100 factors seems like a lot. I bet you are wondering how they scan the environment and develop the catalogue. I was. Then I found out that first they ask the client about what drives his business and the most important factors shaping his business today. Then, the catalogue is built 50/50 with factors from UNITY's point of view and from the client's point of view. Herbst explains, "We have a research team and a lot of these factors are applicable for several topics [this makes me conclude that they are dealing more with macros and megas]. If we talk about the development of the economy or society we already have information that we regularly update. There are people who constantly research the topics but then we try to look at what other people are currently saying about the specific topic". So, UNITY doesn't try to compete with say Deutsche Bank's research on economy, but it compiles the information about the future that is relevant for their customers, on a regular basis. I asked about customers input: "They

are very informed regarding their core business. They know it better...but the future is shaped by different players, by things going on in the market and mostly at its periphery where it is not so visible", says Herbst. This is something that the clients cannot spot without the help of companies like UNITY.

THE LAUNDRY CASE

Example, example, example, please! One of UNITY's clients is a laundry detergent company. They are very good at understanding how their clients are washing and they know all the details about the chemical formulas they use. It would be out of place for somebody else to come and tell them how to proceed in this area. This company would instead prefer to fully understand what is happening in adjacent markets, for example in the textile industry. The client cannot contribute with factors that lie in the periphery of their core competences. This is UNITY's duty. For example, continuing with the case, it is UNITY's job to think about how people are going to wash in the future. After reading the influencing factors, they can presume that, since today everything has become simplified on the machine itself: less buttons, digital display etc, it will have sensors to determine types of fabrics, weight and so on. The future machine will decide by itself what to do. This scenario is following the path of a trend ('minimalism') that we see today.

A contradictory scenario that could come out of a mental exercise similar to the previous one, is 'washing will become more technical' (this relates to the 'back to the roots' trend) and thus we may have many parameters on the washing machine, with you being in charge of getting the desired result. This scenario is somehow contradictory to what is going on today. But we can imagine it as well. We should allow ourselves, during the scenario technique, to ponder on the consequences of this coming true.

A different scenario may sound like this: you can control the machine after each wash, you put the clothes inside and then the washing machine starts the process. Then, you retrieve the result (your washed clothes) and you adjust the washing machine process by telling it you wish for a softer or

a more powerful smell. This would be a completely new way of interacting with your washing machine.
Ok, going back to the process itself, we have the list with the influencing factors, the 80-100 remember? Until this point in the application of the scenario tool, the projections were related to what's going on today and didn't focus on what the future might bring. A system analysis follows in order to find out which of these factors (80-100) are the most important. Here, 'important' means holding an influence over the other factors and spotting the relevant ones for the client, for example the price of energy. Now we can start talking about the future. The influencing factors are the building blocks of the scenarios. UNITY starts thinking about how the future might look if tailored by each of these factors. For example, if we imagine that energy prices will be very high, we have to try to find innovative ways of dealing with the situation, by finding alternative energy sources. That's briefly how the scenario technique works.
If we started describing the future regarding the projections for every key factor (normally there are 10-20 key factors after the pre-selection) then we would have a nice chunk of information on what the future might look like. But that's too much for the client to understand and apply. So, how do you translate this pile of information? UNITY doesn't select them, but clusters them by bringing together the scenarios that intuitively come out of the influencing factors, without thinking about the probability of something actually happening. The clustered scenarios need to make sense together. **The clustering process involves not only thinking about the trend (aka influencing factor) in itself but also what it means for the brand you are working with and what changes that trend could bring in the lifespan of the brand.**
Let's go back to the energy price example. If we talk about urban mobility we can imagine the following scenario: people spend more and more time at home. Everything, including work, is done via communication channels, especially the Internet. We can combine this scenario with another scenario: energy is expensive. It makes sense, right? People spend more time at home

and use energy to communicate instead of taking the train, the car or the bus to get to the location. Less *commuting*, more *communication*.
UNITY developed a method that allows a consistent and easy to comprehend combination of these projections. The end result for the client management is all this information combined in scenarios presented in an easy to understand story, the story of the future.
To reiterate, the delivered scenarios are all very different from each other but that's the whole point. The method UNITY is using is **cluster analysis** and **consistency analysis**. Sounds familiar? It should! Remember, this means UNITY is going through each scenario checking to see whether they fit or not with the situation of their client. I'm thinking, wow, how many nights are they staying up in a row in order to analyze this unimaginable number of scenarios? Just think, if you have 20 key factors and 3 projections for each of the key factors this would generate a matrix of 60 by 60 lines. Herbst comforted me by breaking the news that they are not all alone in this. They discuss the projections with the client and then apply a mathematical algorithm to calculate the scenarios. The aim is to give the most different and unique scenarios possible. My first reaction is 'ah, I'm relieved that there is an algorithm for this.' Then, my second reaction is 'hmmm...I would like to understand how such a subjective (somehow creative) process, is crunched by a mathematical system.' Actually, come to think of it, I don't believe in the power of algorithms and probabilities but hey, I can't deny that UNITY knows its stuff. The result is usually three to five final scenarios. There are millions of possible combinations so they don't know how many will come out after the calculations.
So, we've got to the core meaning of scenarios now, what's next?

WHAT COMES AFTER BUILDING THE SCENARIOS?

UNITY's job doesn't end here; it just starts with scenarios. The key question is what you do with the scenarios and how you use them for planning long-term strategies. Of course, the scenarios are useless if you don't answer relevant future questions with them. Scenarios are just a means to an end", says Herbst.

UNITY assists the customer in different ways and depending on each customer's questions, they either:

1. Have a specific question regarding the future such as: Where should we put the R&D money? Is our strategy secured in the future? Or what kind of fields should we try to innovate in the coming years? This is where future knowledge is used by UNITY with their clients. "Customers usually come to us when they have a fear about the future or they see that something is happening on the market – new technologies arise, new players in the market etc. They usually want to know what would be the future in this or that", Herbst adds. Then, UNITY's task is to spot the underlying question behind the verbalized ones. They try to get under their clients' skin and feel the real issue to elevate opportunity.
2. Express their concerns such as: We have a lot of ideas but don't know which one to focus on, we don't know what's more important for us and where we should invest our budget or we have/don't have a strategy and we want to know about the market of the future so we can derive a strategy which is future oriented. "A lot of companies say, 'if I knew what the future of my customer looks like, how they would behave, I would know today what to prepare for", says Herbst. The big question is: How do we understand the customer of the future?

For each of these ideas one needs to research, develop and find a solution. Scenarios can help in deciding which fields of work are interesting for the future. From there, one can develop ideas around these search fields or innovation fields. This makes innovation more effective!
But you don't need scenarios in every case. You can also just spot trends and consult accordingly. Remember, that's what CScout does. I was then thinking why would a company choose one approach over the other? For example: megatrends, or scenario technique? Megatrends are something 100% definite and can help you understand the future: talking about demographics you can see how society will develop in ten years time. Herbst gets into the

specifics, "When using only megatrends for future strategies, we neglect the aspect of uncertainty. When using scenarios, the uncertainty aspect comes in and allows you to include developments that are not covered by long lasting trends." So, does it really depend on the courageousness of the company? Herbst sighs and answers: "Imagine you are a company that builds washing machines. You spot the trends: miniaturization, individualization and then you tap into these trends. But everybody else is doing this and therefore, it wouldn't help you differentiate on the market. You might be innovative but you cannot break through. If you start using scenarios you are allowing future thinking to enter your strategy." This is very valuable for a long-term perspective. Think about it; it's easier to be successful when you don't build only on the common factors and when you take into consideration and prepare for various scenarios. "Many decision makers want to bet on one future scenario, falling prey to the seductive temptation of trying to predict the future rather than to entertain multiple futures. Another trap is to take the scenarios too literally as though they were static beacons that map out a fixed future. In actuality, their aim is to bind the future but in a flexible way that permits learning and adjustment as the future unfolds. Once the scenarios are finished, the real work starts of how to craft flexible strategies and appropriate monitoring systems", says Paul Schoemaker in *Profiting from Uncertainty*. This sure made me excited to continue the trip and find out what comes next after scenarios.

// THE FUTURE OF THE SCENARIO TECHNIQUE

There are a lot of barriers in scenario thinking. It represents a technique but mostly a mindset. In order to embrace it, you have to be open to the future. If people spot a development that might not be good for their company, they tend to close their eyes because it's painful to consider. Although the scenario technique has gained much adherence in the innovation industry, its subjective and heuristic nature leaves many academics uncomfortable.

In the innovation management scene you always have the 'not invented here' and 'the idea is not good because we didn't find it ourselves' syndromes. Unfortunately this happens between departments and also between technologies invented by institutes and universities. This type of thinking is completely opposite to open innovation. According to Herbst, "People should be allowed to think and ask questions that are painful. On the other hand you have to think about what is good behind the change and how can we use this".

The next big thing in innovation management is learning how to be effective and that means answering a set of questions: How do we come up with the right ideas? How do we combine ideas to be more effective? How do we consider future developments in generating new ideas? The way the ideas support each other is most of the time not taken into consideration. "Innovation management and scenarios will be more important in the future. We will see more sophisticated tools. During the last ten years many companies implemented an innovation process, which helps them to be efficient in handling creative ideas. One problem within this process is that the innovative ideas are evaluated in isolation", says Herbst.

"As scenarios become more popular, as digital natives age, bringing in the broader gaze of futures studies will become increasingly important to ensure that scenarios are used to their fullest capacity", says Sohail Inayatullah in his article *Questioning Scenarios* from the *Journal of Future Studies*.

In the same article, award winning fiction writer Rosaleen Love poses an interesting question regarding the future of the scenario technique: "Who will be using the scenario technique ten to twenty years from now? Future decision-makers will have one thing in common that will set them apart from today's groups. They will have grown up playing with scenarios in the various shared worlds of computer, video and online games (and their future incarnations). Scenario technique will not be unusual but a social technology that will be part of the 'collective intelligence' of the future".

CONCLUSIONS

Let's wrap this up; scenarios are useful as they distance us from the present. In any situation, it's normally better to step outside of it in order to see things clearly. They are useful in creating preferred futures but it should be clear that they do not only take into consideration the positive sides of the future (for the company) but rather prepare companies for reacting to future changes in an efficient 'I did my homework' kind of way.

Scenarios are not meant to be secure options of the future but uncertain ones that could serve as an exercise for companies. The outcome might come true or not. But if it does, you are prepared. **The scenario is only one element in a foresight process.** I was reading a very good article the other day in the Design Mind publication from global innovation firm frog, written by Doug Randall in which he states: "I've spent most of my career using scenario technique, an essentially democratic tool that lets a group of people imagine a range of plausible stories for the future. You're not supposed to ask an expert to pick one; you're supposed to prepare for them all. Yet, scenario planning clients have a tendency to view the professional scenario planner as an augur and oracle. Pioneers of this field such as Herman Kahn, Pierre Wack were routinely treated as if they had all the answers, even if they reminded us that no one does...**As individuals we just aren't that smart. But as groups of experts we have tremendous expertise**".

Poly-social groups rule!

Scenarios are decision-making tools and with them companies can rehearse their decisions before being forced to take them for real.

// WILD CARDS

Wild cards are surprises, but not the 'happy birthday' kind of surprises. Let's stop and imagine that a volcano erupts and suddenly planes cannot fly because of a resulting ash cloud (been there, done that). This was a wild card. But if any of the airline companies had worked with wild cards when building their future strategy, maybe there would have been a faster solution to the problem. The reason why I chose to talk about wild cards now is because the term is often used together with scenarios. Initially, wild cards seem to be just different types of scenarios: like mind-blowing, outstanding, high-impact and very low-probability events. There has been a long debate going on for years regarding what wild cards really are. Are they only negative events or can they be positive ones as well? Every professional chooses to give his own definition and uses the technique accordingly. Don't worry, I won't talk about all of them! In my opinion, what is important to know is that there are two main understandings of the topic:

Wild cards as such are signs of what could happen in the future. For now, we should consider wild cards as extreme scenarios of the future.

> **WILD CARDS** AS SUCH ARE SIMILAR TO WEAK SIGNALS
> **WILD CARD SCENARIOS** ARE STORIES OF POSSIBLE SITUATIONS

So, how do we work with wild cards? By creating wild card scenarios. The technique is the same, only the content is different. Generally acknowledged, wild cards are events with high impact and low probability. From here on, we could start talking about what low probability means and whether we usually consider all the uncertain things about the future (everything that is shocking us) as having a low probability of taking place – but I prefer to keep it simple.

Wild cards are sometimes considered similar to weak signals because of their low probability of transforming into something powerful and stable.

It would be illogical to build a strategy on wild cards, because again, you would be betting on something that has a very low probability of happening. They are about having an **emergency plan for the future** and keeping an eye on the evolution of the low probability events that might increase (aka early detection). Imagine how easy it might have been if BP had explored the wild card of an oil spill or if Japan hadn't built a nuclear reactor on a fault line and exercised the tsunami wild card.

A wild card can also be a business driver, an event that builds momentum and triggers an opportunity for your business. For example, in the 1970s Royal Dutch Shell saw the need for more refineries as an opportunity – the driver being the 1973 Oil Shock. Events, like the fall of the Berlin wall are

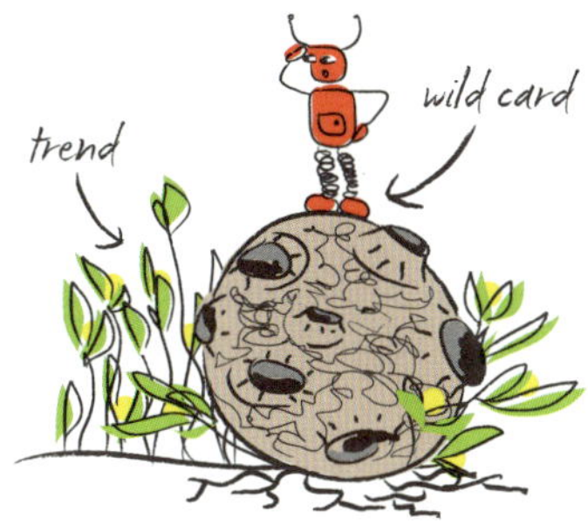

good examples of a wild card. This shows us that the impact can also be positive. As, Michael Herbst from UNITY says, "A lot of people will tend to understand wild cards as having a negative impact. They should also consider the situation when something happens to the company, which dramatically increases the demand of the product. This could be a positive thing!"

Wild cards – events that are unlikely, but that would have enormous consequences – can slow a megatrends' development or create counter-forces. Megatrends say something about the probable future, but there are other possible futures. Every megatrend can be set aside or can suddenly and fundamentally change course of actions. For example, the events of September 11, 2001 temporarily stopped growth and slowed some aspects of globalization. Before it happened, September 11 was a wild card. If such an occurrence takes place so quickly that a usual-planned management process cannot function, the organization will be vulnerable. **Wild cards are unprecedented developments. Therefore, wild cards can break trends.** Thus I agree with Ian Mitroff, regarded as the founder of the discipline of crisis management, when he states, "clearly, it would be the height of folly not to plan for the occurrence of at least one major disaster".

Obviously, this is not easy to deal with or incorporate into future strategy. I mean, let's face it, how many CEOs want to think about what will happen to their company in case of a meteorite hitting the Earth? Up to now research on wild cards has been ad hoc. I was reading *Was It a Wild Card or Just Our Blindness to Gradual Change?* Article 61 from the Journal of Futures Studies, November 2006 by Elina Hiltunen in which she states, "What is clear is that creating wild cards can help an organization be prepared for unexpected situation". It doesn't have to be the exact situation from the wild card. It is again, all about the mindset. It could happen that certain plans from a wild cards analysis are applicable or at least have the potential to create innovation paths when any such random unpredictable event occurs. If we practice we will be prepared for anything. We would know how to start planning and analyzing developments and changes happening all around the world from the beginning if we have the proper mindset. Since wild cards are, by definition, surprises, it might be assumed that there is nothing we can know or do about them. "However, this is not so: not all wild cards are unimaginable", says Hiltunen.

In various articles, wild cards are called disruptive events, structural breaks, discontinuities, surprises, bifurcations, unprecedented developments etc. What is important to remember is that "Wild cards have a direct effect on the human condition, have broad and sometimes fundamental implications, and tend to move too fast for the whole of the system to adjust to the shock", according to John Petersen in *Out of the blue. Wild cards and other big future surprises: how to anticipate and respond to profound change*.
Peterson says in the same book, "In a strategic planning process one can ask three questions with respect to wild cards: which are the most important wild cards for an organization? Can we anticipate their arrival? Is there anything we can do about them? If managers can answer these three questions, wild card analysis increases the possibility that some major future negative events might be averted or moderated, but also positive actions can be implemented".
Petersen's general example of a wild card is a major hurricane devastating a town in a day. He lists almost eighty wild cards (note: referred as scenarios in the back cover of the book) that might happen in the future varying from shift of the earth axis to future prediction becoming a standard business. Futurist.com lists nanotechnology, airplanes that fly themselves, and doubling one's lifespan as examples of wild cards because they are unpredictable and unstable.

This being said, I am out. Next on the travel list is Potsdam, a charming little town close to Berlin. But before I go, I leave you with Hiltunen's Futuropoly game, as promised before. Our brains are now ready to understand all the tools and especially the meaning of the scenarios that she incorporates in the game. Enjoy, and please *do* try this at home!

// *THE FUTUROPOLY GAME*

EXCERPTS TAKEN FROM FUTURETAKES.ORG

We all know learning is not always fun. Sitting in boring lectures that never seem to end, trying to read textbooks that do not inspire or engage you, or trying to learn things by memory without practicing them are teaching methodologies that can make ones enthusiasm vanish. That's why Elina Hiltunen invented Futuropoly, a tool that students can use for learning and testing futures methods themselves. She has been using this method for several years to educate business people about futures tools in the Helsinki School of Economics. The inspiration for Futuropoly was Monopoly. Time has changed this famous game quite a bit; for example, there have been changes in the tokens. In the good old times they included a dog, an iron, and a shoe; today you can find a laptop, hamburger and roller-skate tokens. The monetary system has changed from paper currency to credit cards and the names of the properties have also changed. These changes through the years have inspired a question: What would the Monopoly of the future look like? Well you could only know that if you had the ability to design it yourself.

As a teaching tool, Futuropoly incorporates the basic concepts of futures studies (megatrends, wild cards, weak signals) and various methods (scenario method, futures table, futures wheel, and Delphi method). Some of these methods will be further clarified in the book.

Designing the gameboard is a process of learning how to use weak signals and applying them into practice from the general (macro) level of thinking to practical future business cases.

The first step is to choose the topic of the game. It can be very general (how will the world be in 2050?) or focused on a specific industry (the car industry in 2040) or on a geographic region (Europe in 2050). When Elina was teaching Korean MBA students they used Korea 2028 as a topic for the game. In the Monopoly game, the cards called 'chance' and 'community chest' are replaced with 'wild cards' and 'weak signals' cards in Futuropoly. In her teaching, Elina presents the definition of these concepts and makes the students find the sources of the weak signals and then identify examples of them. In the case of wild cards she encourages students to think about unthinkable and unforeseeable events. The students' task is to identify ten wild and ten weak signals cards. For this exercise, it is important to look for good information sources (the starting point of every futures activity!).

When the cards are ready, participants start thinking about various scenarios (for example in the case of Korea 2028 they sounded like: 'South Korea and North Korea will be united' or 'there will be tensions between them'). Students spend considerable time on this scenario work, because it is a good method for futures work in organizations too. Then, the groups choose the scenario they prefer *and* the one they dislike in order to design the Futuropoly game board.

To create the game board the groups have to think about what successful businesses are going to look like in that specific situation/scenario and represent them on the board. What tokens correspond to future customers and professions? What would be the payment system? (For example, is there money anymore, and why not? Do people pay with credit cards or mobile devices, or is there a return to bartering? What are taxes paid for, and why? What are the types of transportation systems available? How is the game played? Is it a computer game, a hologram or a mobile game etc.?

At the end, all the groups present their outcomes from their Futuropoly game boards. It is amazing how different and innovative their solutions are. The key of this exercise is to give as much freedom to the participants as possible in designing the game board. This inspires them to innovate. Designing while also learning – now that's a teaching method that I can hang with!

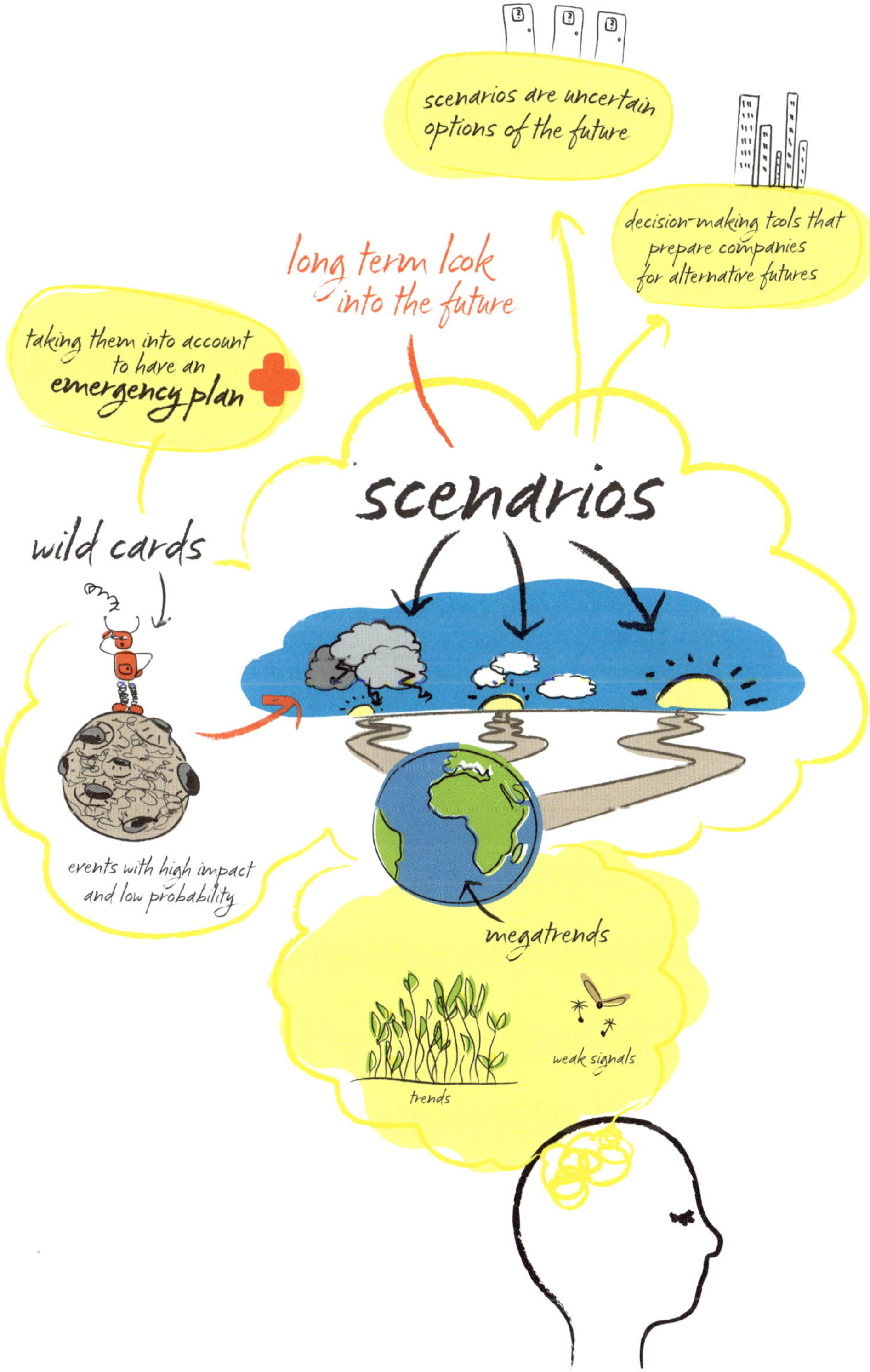
scenarios are uncertain options of the future
decision-making tools that prepare companies for alternative futures
long term look into the future
taking them into account to have an emergency plan
scenarios
wild cards
events with high impact and low probability
megatrends
trends
weak signals

DESIGN THINKING

I knew there are at least three 'doors' to enter the innovations market:

1. **Technology** / e.g. creating touch screens with textures made possible through electro vibrations or other technological breakthroughs in this sense.
2. **Business** / e.g. creating new business models like a financing platform exclusively for grassroots creative projects.
3. **Human** / e.g. asking for desirability. How should a product feel like and behave like for consumers to want to use it?

The most important piece of information to start with here is that design thinking comes from the human approach to innovation.
In his *Harvard Business Review* article, Tim Brown, CEO and president of IDEO – a design and innovation consulting firm – introduced the term of design thinking by giving Thomas Edison as an example of the original design thinker of all time. Brown emphasized the customer-centered, systems thinking approach that Edison used in the creation of the light bulb. After talking to Martin Kruse from the Copenhagen Institute for Futures Studies, this was the second time I had heard about an innovation approach that speaks first about people. We keep forgetting that behind all the innovative projects and products there are people creating for…people.
Some interpret the term 'design thinking' as concerned with the packaging of a product only. This is simply not true. "**Design thinking is a human centered approach to innovation that draws from the designer's toolkit to integrate the needs of people, the possibilities of technology, and the requirements for business success**", says Brown. A growing number of multinational companies started incorporating the approach into their business strategies. Technically, the approach has at its core the designers' traditional way of creating new products (aka problem solving) that is applied to the business world. We are about to find out if this is a happy marriage or not.

Thinking like a designer can transform the way organizations develop products, services, processes and strategy. This approach, which IDEO coined 'design thinking', brings together what is desirable from a human point of view with what is technologically feasible and economically viable for a product to be successful. It also allows people who aren't trained as designers to use creative tools to solve a vast range of challenges.

DESIGN THINKING ON VACCINATION

I found a case study on IDEO's website that helped me fully understand that design is not only about creating good looking chairs. In the design thinking innovation process, design has a special role and its meaning is closer to engineering in a way:

"Iomai, now part of Intercell, approached IDEO to develop a viable method of applying a needle-free vaccine delivery patch. The result is a vaccine patch concept and application prototype that considers the complete user experience, from mail-out packaging to application, and has the potential to make self-application possible for the first time.

IDEO began by investigating a range of skin preparation techniques to enable the vaccine to reach immune system cells in the dermis. By concentrating on the user experience and looking at the needs of patients and nurses in vaccination contexts, IDEO was able to rapidly generate and prototype concepts for application and supporting devices.

IDEO's design is a working model of a concept device meant to ensure consistent application across a range of users. A proprietary mechanism controls the application of the skin preparation and patch location. The patch is shelf stable and has the potential to be sent through the mail and self applied, enabling optimal delivery for pandemic vaccination and large-scale vaccination in emerging economies". If you thought design is only about creating good looking chairs, I am sure the vaccination project conducted by IDEO helped you seeing a new side of what designing good products means.

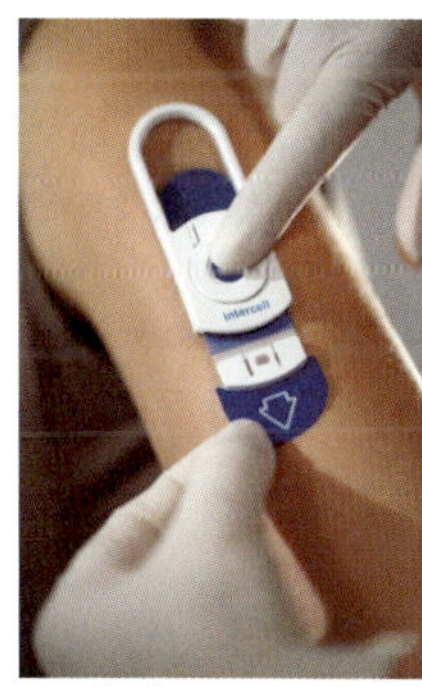

Needle-free vaccine delivery patch / IDEO

// DESIGN THINKING @ POTSDAM D-SCHOOL

Bringing the design thinking approach into the classroom is the ultimate experience. David Kelley, founder and chairman of IDEO, initiated the Institute of Design at Stanford University, California (otherwise known simply as the d.school) in 2004, a playground where students learn how to detect and solve problems and come up with creative solutions through the design thinking approach. The school was built with a donation from the German entrepreneur and co-founder of the software giant SAP, Hasso Plattner. Also thanks to him, there is a Hasso Plattner-Institute D-School in Potsdam, Germany, a chic and charming little town near Berlin where only a certain class of people can afford to buy a house. In my adulthood, there has never been a place where I felt more curious, playful and excited than I did at the D-School in Potsdam. As soon as I entered the door, my heart was pumping as if I had drunk a full glass of liquid energy. I was in the middle of a playground where even the bathroom was inspiring, it had small post-its and pens glued on the inside toilet door. **You know, creativity can strike at any moment!**

Ulrich Weinberg

My interviewee, Professor Ulrich Weinberg, Director of the D-School in Potsdam, was held up in a meeting and I was kindly asked to wait for a while. In any other case I would have been impatient, but at this moment I couldn't have been happier; I could have stayed there for hours without getting bored. I had everything I could wish for at hand: books about innovation and design, Lego, paper, textiles and glue; my time was not lost but *invested*.

I started flipping through the books and writing down titles that I wanted to give a thorough read. Even if I had my laptop with me, **the space pulled me towards non-digital actions** so I started to write my thoughts with a pen on a piece of paper. You know, like the old times. Suddenly, the table and the sofa moved! Yes, everything is on wheels at D-School. I later found out that the environment is supposed to be ever changeable, depending on the needs of the working group.

It is equipped with completely movable and re-designable spaces; you can connect furniture and generate a new work area in seconds. The D-School believes it's better to work in a standing position because it enables you to be more involved and flexible within the process. The only standing collaboration tables readily available were surgery or cocktail ones; hence none of them were suitable so the D-School had the tables built to specification. White boards surround students, a three-dimensional physical information space that waits to be filled with pictures, texts and post-its. The entire space invites you to experiment and is equipped with the type of craft materials that you usually find in a kindergarten. When empty, the space looks like a clinic but in less than an hour everything could be completely covered with colors, images and text if people are working there.

One of David Kelley's beliefs is that "true innovation can only take place when strong multi-disciplinary groups decide to create a common culture and then research the interface of differing opinions and perspectives". This is exactly what the D-Schools are doing, establishing a new culture of innovation by grouping students in small multi-disciplinary teams. In this way, they can uncover unexpected innovations via combining their different points of view on the situation. Imagine students coming from different backgrounds: graphic designers, lawyers, economists, PRs and musicians sitting together and looking at how to solve the same problem. Special people, special place and special processes – **the power of multidisciplinary teams** is truly brilliant! "We think that the complex problems we are confronted with right now need complex solutions and they can't be solved from only one perspective but from very different ones. This is why we put together people from different disciplines in a group of four to six people", says Weinberg.

Surfing on the Potsdam D-School blog, I stumbled upon some very honest opinion coming from an ex-design thinking student, Felix Speiser. He was asked for his definition of design thinking:

"I believe that every single student and teacher at the Hasso-Plattner Institute D-School would give you a different answer to this. For some it might be the process, others emphasize the space or the spirit. To me it is almost a way of life. It's many small learnings and experiences that make you look at the things differently. It means learning that failing has to be done right so you can learn from it. It's understanding and embracing the value of honest feedback and taking in the importance of user insights for product and service development. It means learning to play again and not taking everything dead serious. About being visual, about having fun at work and the importance of building early prototypes of everything. It's also about understanding and valuing the individual strengths of your team members. If each one of us manages to take away and adapt just one of them in other areas like business or healthcare, the impact could be huge."

THE 6 ~~STEPS~~ MODES OF DESIGN THINKING

When I first heard about the ‘6 steps‘ in the design thinking process I almost jumped out of my chair. I learned to be skeptical when it comes to creative processes made in steps because, most of the time, creativity is associated with chaos and unorganized things. ‘10 steps to become a leader’, ‘5 steps to make your boyfriend love you’ are the types of organizational titles that I don’t associate with creativity and also scare me. But Weinberg’s reply made me change my opinion: “If the creatives are individuals and have their own way to generate something new, then, an own created process is fine. But if they start to work in a team, I would always suggest using the design thinking approach”. Ok, so **the 6 steps are about achieving creativity within a team**. Reading Tim Brown’s blog I found this quintessential detail of the approach: “The design thinking process is best thought of as a system of overlapping spaces rather than a sequence of orderly steps. So, I decided to call them ‘modes‘ instead of ‘steps‘, as IDEO likes to say sometimes as well”. Weinberg explained to me the phases of the design thinking process (originally coined by IDEO) by talking about one of D-School’s recent projects that deals with mentally handicapped children. The question they had to answer was: How can we help mentally handicapped children to move around independently in public spaces?

1. Understand. We should first try to understand the problem and we can only achieve this by empathizing with the people involved and by asking them the right questions.

2. Observe. In order to really understand what the needs are, asking the right questions is not enough. Watching what people do and how they interact within their environment gives us clues about what they think and feel. This can be difficult because our minds automatically filter out some information without us even realizing it. We need to learn to see things ‘with a fresh set of eyes‘ and empathizing is what gives us those new eyes. In the disabled kids example, in order to find out how to help them, we have to observe them, accompany them and talk to their

families. Some disabled people can't write so we can't just ask them to fill out a form. Hence, this part of the project takes time, as it offers a 360° view on the problem.
RED ALERT: Does this mean that trends are involved in this phase? Weinberg states that: "We try to free ourselves from all kinds of 'trends' and just focus on the needs of the people we are observing."
3. Define the point of view. A synthesis of all the researched information is made with the goal of finding that very special person for whom we are trying to solve the problem (generically). This could sound like: Mike, aged 17, living in California with his parents and going every morning to a special school for the disabled, accompanied by his sister. This is the hardest part of the design thinking approach, breaking down all the information we have from all sources and observations to only one complex person that is the melting pot of all the problems that have to be solved.

So far, in the first three modes, we shouldn't think of any solutions. This proves to be very difficult as the human mind is trained to do just that. We just have to dive in and learn as much as possible before jumping to conclusions – try to become 'Mike'.
The following three modes bring about the solution.

4. Ideate. Mentally, it represents a process of lateral thinking in terms of concepts and it is a mode of 'flaring' rather than 'focus.' Funnily enough, the ideation mode is the shortest of them all. It normally takes an hour or two and it generates hundreds of ideas that are written down on post-its. Clustering them should deliver one or two main ideas.
5. Prototype. Prototyping the main idea that comes out after the clustering process is maybe the funniest part of the design thinking process. If it's a product, the participants build it with Lego or clay. If it's software, they simulate it on the computer or simply sketch the solution on paper. If it's a service, they role-play it. Why all this? This step is needed in order to start a conversation, to learn or even to fail (if it's the case) quickly and cheaply.

6. Test. Testing is the chance to refine the solutions and make them better. Putting the prototype in the user's hands and watching him using it or misusing it, is the best way to get new insights to further develop it.

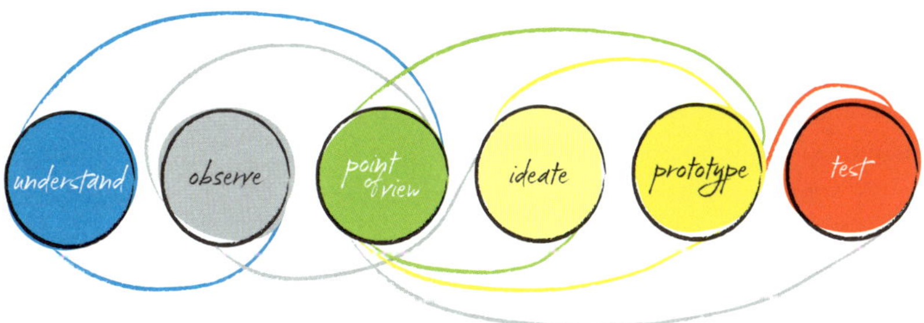

The modes are not steps that come one after the other in a systematic way but can run in a loop, depending on the difficulty of grasping the information for the specific project. Sometimes, with the first prototype, we find out that we were totally wrong or close to being right, but we still need more ideas. It can happen that we have to go back because we misinterpreted or forgot about the legal frame of our project.
As I mentioned above, this is what IDEO coined as a design and innovation process and they have been applying it for over 30 years now. The steps or modes they use are part of every project but may be placed in a different sequence or occurring a higher or lower number of times. In this process you have to be creative when starting each project because there is no written recipe and **every project needs a special approach.**
Design thinking is a systematic approach in the way that it is composed of elements. Most of the time these elements make sense and can be linearly followed but other times you just have to stay open along the process. "You have to stop, go back, iterate, jump over, sometimes you have an inspiration coming from somewhere that you didn't include and doesn't fit in any of your steps but it makes sense. If it makes sense, great!" declares Alexander Grots, ex IDEOer & founder of Gravity – a human centered innovation company.

// DESIGN THINKING VS. USER DRIVEN INNOVATION

Wait a minute – Human-centered innovation? As we just learned that the design thinking process is mainly about placing the human in the center of the process and dedicating all the work to help solve his problems, does this mean human centered innovation is the same as design thinking? Quite frankly, yes. Before finding this out, I was always trying to spot and understand the differences between user driven innovation and design thinking. Both approaches sounded similar and I couldn't feel **the difference between the human aspect and the user aspect.** It took some time before somebody finally lifted the veil. This person was Alexander Grots, who I managed to hunt down between the crowds at TrendONE's LaFutura event in Berlin. He is quite a character: after he worked as a paramedic, founded a software startup that was a client of IDEO, worked in Hong Kong and London in the hotel management business and set up call centers, he then joined IDEO in 2001 and became managing director. In 2008, he decided to found his own company called Gravity together with three other ex IDEO friends. Yes, the typical run-with-the-knowledge and DIY kind of company.
Design thinking and user driven innovation are similar approaches because both of them focus on cross-disciplinary work. In a way, user driven innovation is part of the design thinking approach. Grots explains that **the key difference between design thinking (which is the same as human driven/centered innovation) and user driven innovation stands in the fact that design thinking is not only about the user but also about the: non-user, consumer, shopper, stakeholder etc.** "Every user is a human but not every human is a user. Humans make decisions and fulfill the effect of whatever technology is delivered. So there is always a human involved in the process, no matter what this process is about. Think about moms that buy toys for their kids or men who buy technical equipment for their kitchens where their wives are spending much more time than they do. The user is not always the buyer or the decision maker", says Grots.
At this point I made the connection with something that Weinberg told me during our talk at the D-School, when he gave the example of mentally disabled people: "In the project with mentally disabled people where we

were trying to find out how can we help them to move around independently in the public space, we couldn't call them users. They are just people who have needs. And we try to find solutions for them".

Grots helped me round out the concept of design thinking by telling me an interesting story about how design thinking started as an approach and then became a discipline. "The designers approach and IDEO was founded in the States where design is not only describing forms but also engineering solutions and products. IDEO was a pure engineering company". I am paraphrasing Grots here: Remember the initial Coke bottle? Everything about its shape had an impact on the brand and on the production line. When the bottle was placed for the first time on store shelves it also had a certain impact on people, their behavior, how they were holding and drinking from it. In order to take all these things into account, the designer's job is not only creating the aesthetic features: colors, packaging, forms, but also 'engineering' the whole product.

"Whatever it is that I'm creating should deliver a holistic outcome. Products live in contexts and that's what I have to anticipate. This is why I think the design thinking approach was always connected to common sense but it was unspecialized", says Grots.

Now, design thinking is used like a translator between a designer's approach and the business world. / Alexander Grots

I came to realize design thinking is connected to consumer insights as well – isn't it? But there are companies who only do consumer insights, so what's the extra benefit of design thinking? I get a short and obvious answer from Grots: "As I said before, you don't only include consumers".

How do we know when we are dealing with a brainchild of design thinking and not the result of some other innovation process? You have to ask some questions. Tim Brown says it best on his blog, "In other words, were new choices created not only about the product, service or experience but also about the business goal itself? Did a product become a service? Did a service become an experience? Were entirely new users or markets identified? Were new to the world offerings created? These questions seem to reflect the higher bar that I believe we might expect to be the result of a broader application of the design approach".

// @FROG

Till Grusche

I came to know frog long ago, in a time when I didn't have a flat in Berlin. I was constantly traveling so it was hard to find the time to actually see apartments. There I was, skyping with Till Grusche, Europe Marketing Director of frog, sitting on the floor of a bedroom that was temporarily mine and trying not to unplug the Ethernet cable from my laptop, as there was no Wifi. Stone Age-style.

Anyway, frog is a global innovation firm and they take care of the innovation process from one end to another, meaning they can work for clients from early research to the ideation and design phase and on to the delivery phase as well. After delivering the ideas, concepts and presentations frog can also build the real prototypes or the code in the case of digital products. Placing the human at the center of their innovation process makes them a good candidate for understanding a new point of view on design thinking.

The firm has a three-step process. Very generic, Grusche says, "almost every company that does innovation has kind of the same approach". Modest, I would say. In short, it's about:

> **Discover** – intensive design research and strategic analysis, gaining insight into customers, competitors, client brand, and key opportunities. Discover inspires and guides Design and Deliver
> **Design** – produce design approaches and concepts to fully address the client's challenge
> **Deliver** – guarantee the accurate translation of ideas into reality

The actual creation of the product is essentially what differentiates frog from other companies. One thing you have to imagine is that these steps merge; one does not have to stop for another to begin. This is the big strength of an innovation company like frog that offers the possibility to collaborate at very early stages with the people who generate insights and create ideas. You get a comprehensive view of the situation if you are involved in every step.

But what I was keen on finding out from frog was about the **design oriented consumer research**. This is consumer-focused qualitative research, which means they observe people's actions and how they behave in their natural environment (it almost sounds like we are studying some sort of endangered species) and then translate them into success stories. **It's an ethnographical approach to understanding the reasons behind people's choices**. Or better said, a design thinking approach. It is about understanding behaviors not demographics, looking at markets and contacts more for inspiration than validation or certain assumptions, looking at small audiences rather than large sample groups.

This kind of research is an important addition to quantitative market research. "Numbers, traditional market research helps you have a good overview on the market but in the end people are not numbers and points. It happens very often that you can learn from the extremes. For example, from people who adapt products very quickly [aka lead users] that can give you a good insight into what a product is really about", says Grusche. Before drawing the conclusions, frog combines this type of research with information about trends. "Trends influence many of the requests that our clients have. It is obvious that if you have a trend called social media then this will play a role in what our clients think about their new product. So, trends represent the base. Trends also influence the way frog designers think and evaluate certain things out there from the market. But they are not building innovation only based on jumping on a trend", says Grusche. At frog, everything is thought through over and over again.

TUPPERWARE FLAT OUT

frog managed to transform insights for Tupperware (people having troubles finding the lids and depositing the cases) into a coherent solution. The Tupperware case is not among the latest success stories frog has come up with, but I liked its use of engineering in design thinking; even if it's purely an industrial design example it demonstrates well the notion of insight - to - solution. Today, the company does a lot of convergent design work and software development. The old 'industrial design' approach isn't the core of their existence but hey, it's the frog heritage and it still plays an important role.

“Ever the innovator, Tupperware sought to improve the design of their famous product for consumers who need more flexibility in their storage. Problem: Many people devote an entire cabinet to storage bins, yet they have trouble finding a container that is the right size. And when they do, they have trouble finding the lid that fits. Tupperware wanted to help users consolidate and quickly find matching lids and bases. They approached frog for this task.

To optimize storage flexibility, the Tupperware containers needed to be malleable. Everything that transformed itself was included for study. A diversity of inspirations like the Chinese takeout box, origami, bendable straws, and spring-loaded mechanisms were explored. Many of the ideas were rough-prototyped, from the most whimsical to the strictly mechanical. The winning concept was derived from the accordion-style expansion of drainage pipes. frog created a form that would flatten to occupy little storage space and then extend into a full container when in use. These can even be loaded into the dishwasher like plates and washed flat. The new design is also color-coded, so it’s easy to locate the right size container and its corresponding lid; frog delivered design prototypes and design-for-manufacture (DFM) direction straight to Tupperware for engineering intent. The end product is the efficient, flexible FlatOut!™ Tupperware series – an overwhelming success.”

FlatOut!™
Tupperware
/ frog

// THE FUTURE DESIGN THINKING

The growth from 7000 Google hits on design thinking in 2007 to the 400.000 today makes Weinberg optimistic regarding the development of this innovation process that brings together the design and the business world: "I wouldn't call it a happy marriage between the two worlds yet because we still have a lot to do with the people who don't trust the approach. There are people that are open but it always takes a lot of time inside an organization to rethink how it is run, where to focus and how to come up with better, more sustainable solutions. This change is something against the normal procedure of a company so it will take several years until we can really say there *is* a marriage".

In *The Ten Faces of Innovation,* Tom Kelley acknowledges that it's hard to admit but "by nature, experts are often the guardians of conventional wisdom". Along the same lines, Grots agrees that there are people who get stuck and protect themselves with the knowledge they have and processes they've already learned: "New innovation processes sometimes are looked at as threats and managers don't feel comfortable with insecurity. If you decide to work with design thinking you have to be at least a bit comfortable with stating 'maybe I don't know what's coming next'. It's about innovation, it's

in the new so that's how it's supposed to feel: uneasy. To support his point of view, I remember when I attended Eurobest 2010 – the European Advertising festival – in Hamburg and AKQA's (a new breed of agency with innovation at its core) Co-founder and CEO, James Hilton, said something that hit me: "Creativity is fear. If you aren't shitting yourself, you're not doing something new". Is that how creativity is supposed to feel?! It's no wonder that some of our most creative minds are also those who enjoy risks.
The design thinking approach has become more and more appealing and the current crisis has helped add to this. 'You hear top managers saying that the old way of seeing future developments – by extrapolating the present into the future and doing quantitative market research (the databases type of research, where every person was actually just a number) – worked but now something has changed.' The hard facts (aka the numbers) are of use in some aspects but the soft ones (qualitative) have a much more important role for the future. Managers don't know how to work with these soft facts yet and this makes them open up to the trends and futures industry. They want to know how to work with trends and with things that are happening *now* even if they can't grab them – as they are not numbers, but simple facts.
In terms of scientific further developments of design thinking, in addition to the D-School in Potsdam, the Hasso Plattner Institute is conducting a research program on 'team based innovation', both in Germany and in the USA. Thirty researchers are trying to find out why it works, how to improve it and where can design thinking be applied. There is certainly more to come from this research program, just keep an eye out for it!

Even though I'll never bore of this favorite topic of mine, I am going to leave it for now. We know how to work with ideas, deal with change and contextualize information; we should take a break and reflect on what has been presented / discussed up until now. I am taking my break somewhere in the middle of Germany close to the woods where Hansel and Gretel – the fairytale – was conceived. I hope to find a house made of candy; after all, today is my birthday.

design thinking
needs of people
technological developments
human-centered approach to innovation
requirements for business success
cross-disciplinary work
detect and solve problems
create a common culture
engineering solutions that have a holistic outcome

HOW TO BUILD STRATEGIC FUTURES?

Transformation

STEPPING INTO STRATEGIC STUFF

In order to figure things out in the following chapters I needed to take a paper and a pen, sink myself into total silence at home and start spotting the core differences and relevant meanings regarding future strategy processes. I shut down all the instant messengers I had opened on my computer and turned my phone off, then started to digest the approaches I had previously written on post-its. After clustering them, I narrowed down all the processes I've learned about, to the three most important topics regarding long-term future strategies and planning, that I felt needed a place in this book. Tada! Here you have it: forecasting, strategic foresight and strategic planning. Sure, there are many more future studies approaches but to make it less complicated I'm going to stick to these three.

All the processes deal with long-run plans and strategies. Planning can also be done for shorter periods like one to three years but forecasting and foresight are in the game for five to 50 or even 100 years time span. Therefore, the planning we are going to talk about includes the results coming from these two processes and so it becomes strategic – strategic planning.

BRIEFLY

> **Forecasting is about predicting one future (one scenario) from known current trends by making estimations and looking at quantitative research.** (e.g. In 2050 80% of the population will be over 60 years old).

> **Foresight is identifying possible future states (alternative future scenarios) for which sometimes no previous data is available.** Now imagine a capital T. We could consider foresight as the horizontal line of the T, representing lateral thinking and forecasting would be the vertical line of the T, which helps to deepen the knowledge on a selected future scenario, determined through the foresight process. (e.g. In 2050, the work force might drop/decrease).

> **Strategic planning deals with the preferred future and involves the overall set of methods and tools for taking decisions on how to reach these 'future states' that are identified as relevant through forecasting, foresight or other methods.** (e.g. To prevent the workforce from decreasing, we need to start creating job opportunities and assignments for people who are aged over 60. To achieve this goal, we need x, y, z.).

I was reading an article entitled *From Forecasting to Foresight Processes – New Participative Foresight Activities in Germany* from the *Journal of Forecasting 22/2003* published online by Wiley InterScience and written by Dr. Kerstin Cuhls – scientific project manager at the Fraunhofer Institute for Systems and Innovation Research in Germany that instantly revealed my previous thought process. I was sitting down and waiting, waiting patiently for everything to be unmasked. And it did, it rolled out in ecstasy at my feet (as Kafka promised). Suddenly, the curtain was pulled away and the world of futures just offered itself in front of my eyes.

Well, not really, but here's what I learned from Cuhls: the definitions of our three topics "vary to a certain extent, but they all have the view into the future in common. The future is unknown, but the broad, general directions can be guessed at and reasonably dealt with. Foresight goes further than forecasting, including aspects of networking (which is similar to what 'open innovation' means today) and the preparation of decisions concerning the future. This is one reason why, in the 1990s, when foresight became popular on a national scale in many countries, the wording also changed from forecasting to foresight. What does the result of a futures study mean for the present? Foresight is not planning, but foresight results provide 'information' about the future and are therefore one step in the planning and preparation of decisions…yet the differences are relevant for the development of theory, concepts and methods". The funny thing (or not) is that, many authors working with these processes do not distinguish between foresight, forecasting and planning.

Foresight
not only looks into the future by using all instruments of futures research, but also includes utilizing implementations for the present (meaning, information about the present such as trends).

Kerstin Cuhls

I knew that if I really want to get to the bottom of these concepts I had to get in touch with Cuhls and discuss the topics. Before contacting her I had butterflies in my stomach as if I were about to speak to the author of a childhood book that marked my entire existence, but she quickly made me feel at ease.

"In forecasting you already know what you are looking at but in foresight you are searching for what you have to look at. We apply both methods at the Fraunhofer Institute and we are still struggling to combine them more efficiently. This is difficult especially with bringing the quantitative forecasting tools to the foresight perspective", says Cuhls during our talk.

Speaking of quantitative, (you are already familiar with the current perspective of numeric calculations in this book), I was wondering if the whole forecasting process is considered outdated. Cuhls clarifies this issue: "Forecasting being outdated is an existing point of view but I would agree with it only for the forecasting process that still pretends it can *predict* the future. The forecasting process that delivers data as a working material (projections, extrapolations) to integrate in a specific context in the foresight process is useful and still makes sense".

The switch from understanding forecasting as being a prediction tool to merely a data finder, was made at the beginning of the 1990s, when people finally noticed that predictions and prognosis (basically stating what the future will look like) was not possible. Even so, today, there are many companies that still use the old approach and there are still many people who believe in their data. These companies have fixed forecasts and they believe the future will look like they predicted without actually conducting a continuous investigation. It's an 'only plan A, no plan B' kind of mindset.

As time went on, the wording for the processes changed. It happens too often that the novelty of an approach is acquired just by renaming an old concept. "The discussion is the same as ten years ago. Nothing changes regarding terminologies. At the European level, when it comes to future studies methods, they use words like 'forward looking activities' or 'future thinking' when talking about long-term strategy processes. Nevertheless, the terms overlap so much that companies started to use only one of them. Sometimes, a big foresight project integrates the 'job' of forecasting without mentioning it. Especially in USA, both concepts – forecasting and foresight – are considered as the same", says Cuhls.

The fact that there are multiple sides to the same story doesn't only come from subjective opinions and empirical approaches (which most of the processes and tools presented in this book have) but also from the scientific vs. business differentiation of the matter. **So, I couldn't help but wonder if there is a point where business stops and science starts.** "From the use of foresight there isn't any because both sides use the same methodologies. On the other side you can find more trends, forecasting knowledge and creativity methods used on the business side (like forecasting fashion) and you can find more technology topics and hard core market indicators (highs, lows, changes of a specific industry) on the scientific approach. Of course, creativity methods are not acknowledged in science as such. You can use them in workshops to cluster, discuss topics but they are not acknowledged as scientific, rigorous methods. They are just like playing around in science. Nevertheless, recently, more and more scientists have accepted these creative techniques in workshops because they are fun. So, there is no real differentiation between the methods", says Cuhls. **What the scientific world is doing additionally is further development, refining and improving the methods itself.**

While discussing the same topic with Jasper Grosskurth, author of *Futures of Technology in Africa*, I discovered a slightly different point of view. Grosskurth states that the difference between the business and scientific worlds in regards to innovation stands also in the time horizons in which they work: "Businesses need to respond to much smaller changes in consumer behavior. The government doesn't have to differentiate itself from itself – it just needs to be good. Companies that delivered a good product (and not exceptional) don't survive very long".

Racking my brain in the search for more about the scientific perspective, I remembered Grosskurth had told me about The Millennium Project, an independent, non-profit, global, participatory futures research think tank that works for international organizations, governments, corporations, NGOs and universities. I rushed to my computer and started browsing through their *Futures Research Methodology* study, one of the think tank's constantly updated outcomes about – yes, exactly our topic – future tools and methodologies. In the introduction of this study I found something that may help shed some light on these futures processes: Scientifically speaking, "studying the future is not simply economic projections or sociological analysis or technological forecasting, but a multi-disciplinary examination of change in all major areas of life to find the interacting dynamics that are creating the next age", says Jerome C. Glenn – the co-founder and director of The Millennium Project. In a nutshell, "the purpose of futures methodology is to systematically explore, create, and test both possible and desirable futures to improve decisions. Businesses use futures methods to enhance understanding of future markets. The ability to anticipate gives extra time to better understand threats and opportunities, develop more creative strategies, create new product opportunities, and create and share vision for organizational change", states Glenn in the same study. Future research in business is less about accuracy (accuracy in the sense that the future is unknown, hence we cannot accurately predict it) and more about usefulness (the exercise in itself) in planning the future of a company. It's about gaining time to think about your decisions.

Somebody asked me a while ago how important it is to define what all these terms mean as long as you're doing the right thing. But let's imagine you just got a managing position in a big company and you have to start thinking about the future of the company you are leading. You are not an expert so you need to hire one. Where do you start? How do you know what to focus on if these names are not clear? When your computer breaks down you know that you have to call an IT company. Who do you call when you want to talk about the future growth of your company?

Speaking of processes to manage change...prepare yourselves for the future studies chapter of our road trip. I must seriously ask you to switch your brain ON – this is no cakewalk. Palo Alto, Eltville and Hamburg are our last stops, before I go home for Christmas.

FORECASTING

Something that probably everybody understands when it comes to forecasting is the weather. Sunday, 10° C, with a small chance of rain... you know what I mean. Sometimes even fashion forecasting can make sense to the general public. An Answers.com search returns an immediate answer: "Fashion forecasting is a vital activity where experienced fashion analysts scrutinize past buying patterns signs for current market dynamics and project these into the future. There are forecast agencies that collect facts pertaining to consumer buying patterns, activities and occasions for which clothes are needed, their priorities, reading habits etc. Without well-judged projections of designs, colors and fabrics by the forecasters, the producers would not be able to sell their goods; the importance of the forecaster in trend analysis becomes invaluable". Fair enough.

But when words like 'technological', 'economical' and 'product' are added to 'forecasting', things become blurry. Moreover, seeing some websites forecasting 'earthquake predictions' makes things even more confusing.

According to the Millennium Project, experts in the field of futures and innovation differ between exploratory forecasting and normative forecasting.

Note the part about 'how to reach the goals' is considered in this book as strategic planning business.

> Exploratory approaches generally deal with questions of what may, might, or could possibly happen on the basis of the forces at play.

> Normative forecasting almost always reflects the needs of an organization and, therefore, is goal-oriented. Furthermore, the normative forecast consists of two essential parts: the statement of a goal or set of goals for a specific time and the analysis in detail of how to reach the goal(s).

You should notice here that the type of forecasting we are interested in – mentioned before as 'new school' forecasting – is *normative* forecasting because the process blends very well with foresight and offers data to support the process.

"The methods and techniques of normative and exploratory forecasting overlap. [Surprise, surprise] Exploratory forecasts, however, tend to rely much more strongly on mathematical analysis and formal, quantitative trend

forecasting, as well as extensive use of probabilistic methods. Normative forecasting tends to rely more heavily on qualitative tools, since it involves more open-ended, uncertain, and creative elements of the futures enterprise. Many techniques tend to be used in both normative and exploratory formats", says Joseph F. Coates in the Normative Forecasting chapter of *Futures Research Methodology*.

Carefully analyzing all this theory makes me feel that I am reading about the evolution of the forecasting method from plain prediction towards what foresight means. It could happen that we are witnessing part of the development as we speak so let's keep going.

Forecasting does not predict the future in general. In 1983, the *Webster Encyclopedia* defined the verb 'to forecast' as "to calculate or predict (some future event or condition) usually as a result of rational study and analysis of available pertinent data". So, it is focused on *something*. That something is determined by foresight but we will get there in the next chapter. "In forecasting, the area has to be observed or the research questions have to be known in advance. Forecasting is the estimation of the future in a specific research area or according to the questions posed by means of scientific methodology", says Cuhls in her article *From Forecasting to Foresight Processes – New Participative Foresight Activities in Germany*.

Joseph Paul Martino in his book *Technological Forecasting for Decision Making* characterizes a forecast as having **four elements**: "the **time of the forecast** [in 10, 30, 50, 100 years], the **technology being forecasted** [what exactly will appear/happen], a statement of the **characteristics of the technology** [what will this technology do, change], and a statement of the **probability associated with the forecast,** [this will make people consume x% less]".

Graham T.T. Molitor, in his article *Scenarios: Worth the Effort?* from the *Journal of Futures Studies* in February 2009, says that almost without exception, "every study I ever conducted (whether involving hard or soft sciences) easily traces back in time and cross-culturally at least 100 years or longer. The resulting timeline reveals the ongoing direction and pace of oncoming developments. Decades of such research reveal that roots trace back much further – in some

It is always important to know the roots of the phenomena, movement and trend that you are talking about. It makes you understand its essence better.

cases millions, even billions of years, depending on the topic". In the same article he also calls fundamental "Down to Earth" forecasting as being the "pick and shovel" research method. "I've always focused on digging deeply and broadly to gather a panorama of past, present and prospective facts and then couple that with what acknowledged experts have to say". This leads us to the conclusion that older methods that look into the past shouldn't be considered outdated. They can just be inserted into new future thinking processes like foresight for example.

While I was studying Coolhunting at IED, Maya McCarthy gave an interesting lecture about the History of Fashion. "Why do we have to know what happened in the past?" I asked myself. Apart from the fact that it was simply interesting, it helped us better understand current and future trends and the social, cultural, historical, economical aspect that lies behind them. Pero Mićić, founder of the FutureManagementGroup wrote in *The Future Radar* that "not knowing the history makes it difficult to understand the future. For example, in order to understand cultural trends one has to know a bit of the history of cultural movements, how were these created, what were the driving forces and what made them become movements. We often recognize the future based on a pattern from the past".

Grosskurth, in his projects, needs to look at least as far into the past as he looks into the future. He explains: "In a typical workshop situation if we want to make a 30 year scenario on the future of the rural areas in Flanders we invite people to participate and ask them to bring pictures of the last 30 years from that area. After this, we put the pictures in a timeline and we show them the differences from the past 30 years. This way, the group gets a feeling of the 'size' of the change and suddenly people realize how many things were different in the past". Only then he asks them to think ahead. How had this and that changed in the past 30 years and what is going to change in the future? Grosskurth builds up the complexity of the participants' answers by showing them what happened in the past.

Wrapping up the meaning and usefulness of the terminologies, Cuhls says: "There is no right and wrong. The future will unveil what is right!"

// INSTITUTE FOR THE FUTURE @PALO ALTO

Lyn Jeffery

How do you know when a hipster has reached his peak? When you see one on the Berlin metro....reading a book entitled *What Was The Hipster?* An *n+1* research branch small book series #3. No relation to our topic but that was my morning anecdote from the metro ride to the office, where I was going to have a Skype conversation with Lyn Jeffery from the Institute for the Future in Palo Alto, California; the land of innovation. I felt privileged that Jeffery accepted my invitation to talk. She is a cultural anthropologist whose research focuses on the changing nature of identity and reality in a blended digital physical world, with special interests in real-time data and the Chinese-language Internet. Fascinating!

The Institute for the Future started as a spin-off of the RAND Corporation in the 1960s. The company changed the meaning of Delphi from being just a mythical oracle to a future thinking tool. *An Experimental Study of Group Opinion* by Norman Crolee Dalkey on RAND's website tells the story: "RAND developed the Delphi method in the 1950s, originally to forecast the impact of technology on warfare. The method entails a group of experts who anonymously reply to questionnaires and subsequently receive feedback in the form of a statistical representation of the 'group response,' after which the process repeats itself [Therefore, the experts answer from the second round under the influence of their colleagues' opinions]. The goal is to reduce the range of responses and arrive at something closer to expert consensus. The Delphi-method has been widely adopted and is still in use today".

The Delphi-method is especially useful for long-range forecasting (20-30 years), most often in the technological fields, as expert opinions are the only source of information available. Cuhls points out that, "for complex themes it is better to use other methodologies like scenarios and to take into account what Delphi results can provide as single information pieces". In other words: use forecasting only for getting deep-additional data for the foresight process, not as a standalone process. The Delphi-method, various forms of expert group meetings or interviews, statistics and extrapolation, numbers, growth curves, more quantitative than qualitative tools, builds up a forecast. The methodology is not fixed so a mixture of tools is usually implemented.

The Institute for the Future (IFTF) is an independent, nonprofit research group with 40 years of forecasting experience; according to their website: "IFTF focuses on identifying emerging trends and discontinuities that will transform global society and the global marketplace". Throughout the 1970s the US government funded the institute. When Reagan came to power in the 1980s, he extensively cut funding for research, causing IFTF to shift to corporate clients from various industries like health, food, technology and consumer goods. In the last five years, IFTF has made a big push to rebalance their funding so now they work with clients like the Rockefeller Foundation and government agencies like NATO.

Ten years ago, when Jeffery came on as a qualitative researcher helping IFTF understand how scientists were changing their way of working because of the digitalization trend, she knew nothing about forecasting in particular. However, she was finishing her doctoral thesis in cultural anthropology, which made her the perfect candidate for becoming the successful forecaster that she is today. I noticed that most of trend researchers or forecasters come from communication and design studies but always have an anthropological or psychological component embedded. After all, it is about understanding people.

I was amazed to find out that all IFTF's research goes public after one year. This is truly beneficial as it helps the understanding of forecasting to develop, but many of their pioneering projects get buried because nobody is PR-ing them.

Only 30 people currently work at the Institute and they usually tend not to expand to more than 35. I was wondering, why is this the magic number? "We found out that a core of 30, 35 people work well together. We are a networked organization. **One of the challenges for doing forecasting is that you need so many different kinds of expertise and a lot of them are about building and maintaining relationships"**, says Jeffery.

What is special about the IFTF is that they constantly innovate and experiment with new methodologies. "We are methodologically agnostic. We don't define ourselves by a method rather by a very diverse set of researchers and it is important that we're getting people with very different perspectives", says Jeffery. One of the co-founders of the Institute, a German American logician and futurologist, Olaf Helmer, was the originator of the Delphi method. However, from the early 1970s the Institute brought together social scientists with engineers to apply a mixed method approach to understanding the Internet as it unfolded. It's always depending on the people involved. "It's about creating foresight but this is only as good as the process by which you explore the foresight for the people to whom it matters. We spend a lot of time on finding the right technique for extracting the insight out of foresight", says Jeffery.

Futures studies does not *own* such methodologies, but rather *borrows* them from other social sciences: ethnographic and sociological methods, gaming methodology; game mechanics are becoming more important for IFTF. "Consultancy companies create 'steps' for their innovation approach because they are selling a process. We are entirely bespoke. We are bringing all the tools to the table and including interesting ways of communicating the foresight: Prezi, PowerPoint, maps..." explains Jeffery. It depends on who the client is and what question he is trying to answer. If the client is interested in how people are going to use mobile technology in 2014 than it is obvious that talking to real people is necessary. Jeffery believes "It's an attempt to broaden the voices that speak in the process of generating foresight. It's about hearing new opinions, ideas and working with the people involved to come up with new solutions".

The online engaging platforms are another tool for creating conversations among strangers, networks, helping people build on one another's ideas, create something that didn't exist but it's all in response to a proposed topic. So, IFTF is doing normative forecasting (using qualitative methods) and includes everything in the foresight process that is repeatedly mentioned during my talk to Jeffery. As I said before, in the USA they don't explicitly say where forecast ends and foresight begins but IFTF uses both in their projects. The names may differ, but the essential nature of the process remains the same.

In order to get to the insights, IFTF does live-time recording, mapping, scanning, reading and trend research. They are constantly creating new content and materials that can be implemented into future projects. "Others would stick to what they've done and continue with it but because we are doing so much new research, we sometimes move too quickly, we have so much content that someone could be using it to generate an ongoing trend", says Jeffery.

My mind was still blocked by the first information I ever received about forecasting being a linear process that extrapolates data from the past in order to build the future. I asked Jeffery if they use data from the past and she answers: "Absolutely. IFTF regularly returns to its more than 40 years of forecasting to mine previous work for insights. Often, these forecasts are frameworks for thinking about the future, about human practices and technological developments". Can games be framemarks for thinking about the future?

GAMING THE FUTURE

According to Norbert Hillinger, expert in Gaming & Entertainment trends at TrendONE, "Gaming is set to become a driving force across all the generations. But besides having fun, the more serious games show the real world as it is and the aim is to link the action to making a serious point about society. Situations are simulated to players, which focus on collaborative activities and not just high scores and competition".

IFTF has developed a few outstanding 'serious games' in the past few years. Program director Kathi Vian, blogger and futurist Jamais Cascio, and Director of Games Research Jane McGonigal of the Ten-Year Forecast team, developed the Superstruct game (explained below). Listening to McGonigal (an extraordinary character I first stumbled upon on TED.com) talk about gaming as a means to make a better world was an eye-opener. The thought that we can save ourselves from problems of the future by making games today that endorse such problems, is innovative to say the least. "Games like World of Warcraft give players the means to save worlds, and incentive to learn the habits of heroes. What if we could harness this gamer power to solve real-world problems? By spending all this time playing games we are actually changing what we are capable of as human beings, we are evolving to be more collaborative", says McGonigal in her TED talk – *Gaming Can Make A Better World*. Your mandatory homework assignment is to Youtube it; you can thank me later.

EXPLORE THE WORLD OF SUPERSTRUCT

SUPERSTRUCT: [VERB] TO EXTEND, TO BUILD UPON, TO REACH BEYOND PRESENT LIMITS.

"It's the summer of 2019. Imagine you're already there, and tell us a little bit about your future self. Where are you having dinner tonight?" That is how Superstruct, the massive multiplayer forecasting game, enticed people to play. I will paraphrase from IFTF's blog: Superstructing means reinventing our tools and processes, our organizational structures, and even our concepts of cooperation and collaboration. The game was played by more than 7000 citizen future-forecasters from September to November 2008. Although the game is no longer live, we can still learn from its mechanism as a forecasting tool.

Everyone could have played the game! The more players, the better the collective forecast. The game was played on forums, blogs, videos, Wikis. There were five challenges presented, unique and intertwined, named 'Superthreats': Quarantine, Ravenous, Outlaw Planet, Generation Exile and Power Struggle.

Set in the grim future of 2019, Superstruct begins with the findings of a fictional supercomputer that, after a yearlong analysis, predicts the extinction of the human population in 23 years. Potentially terminal combinations of the five 'Superthreats', which represent the collision of environmental, economic, and social risks, are causing massive disruptions of global society. The existing structures of human civilization – from families and language to corporate society and technological infrastructures – just aren't enough to survive.

Playing the game allowed participants to superstruct the institutions and activities of daily life. Six-weeks of global game play helped IFTF chronicle the world of 2019 where players had to imagine how to solve the problems the world will face and to document and record how they, their families, local communities, or extended social networks might respond to a catastrophic population collapse. It was about making the future, inventing new ways to organize the human race and augmenting our collective human potential. "Superstruct is mass collaboration wrapped in a high-stakes, challenging game environment that will appeal to gamers, but also to creative thinkers of all ages and types. By harnessing the thinking and input of thousands, or millions of minds, research shows that we can tackle massive problems faster and find tenable solutions", expounds McGonigal in her TED speech.

"IFTF has always been interested in participatory forecasting – in bringing people together to imagine and even invent the future", says Vian. "With massively multiplayer forecasting games, we suddenly have thousands of people turning their attention to the big challenges of the coming decades. All those eyes on the future will give us a view that none of us individually could imagine".

The game generated many new ideas that could actually be used for a better global future. I'd say that's a hell of an interesting 'Delphi'.

Another interesting example of an IFTF project was done for the Office of Naval Research-US Navy. They wanted to play a game on the future of piracy since they were dealing with the pirates issue in Somalia and so they asked IFTF to build a 'war gaming' – an online platform, pure game designed scenarios and mechanics. The idea is to bring in people who can respond to provocative forecasts and help generate responses to the problem. Game designers, video producers and all those people become part of IFTF to deliver the experience. The content of the game, the process of producing such a game is creating a scenario. IFTF created a robust scenario about the future of Somalia that takes into consideration economic, cultural and international fishing aspects that players can solve.

I'm hoping that all the terminologies didn't set your brain on fire, so we can proceed with strategic foresight and a bunch of cool companies that helped me put things together. Next we're flying to Frankfurt, taking a train to Wiesbaden and a taxi to Eltville am Rhein in the middle of nowhere (aka German fields), location of the headquarters of one of the most innovative agencies working with strategic foresight – FutureManagementGroup. But first, some theory.

one scenario
quantitative research
trends
forecasting
goal-oriented
should deliver data as working material
the research questions have to be known in advance
should not try to predict the future

STRATEGIC FORESIGHT

I can't believe it; I'm trapped again on the train but now it's big time! I was on my way to Hamburg to attend a meeting and all of a sudden Deutsche Bahn, the national railway station, decided to go on strike while we were in the middle of nowhere. These kind of decisions are not typical for Germans but, I guess, as one of my friends once said, "Germans still wish to steal some of the 'Frenchie's' decadence". I found this very funny – at the time. No taxis, no buses, no trains traveling in any direction. Thank God, the ICE trains have power plugs so I could sit quietly for the remaining 4 hours and chew on my topic: strategic foresight.

Everybody agrees that **foresight is a systematic process of identifying futures**. Not just one version of the future, but several possible ways of developing the future. This however, is pretty much the only thing that experts agree on regarding foresight. As we already know, foresight is needed to select the areas that should be examined in more detail via forecasting methods. Strategic foresight is not a linear approach; it's about grasping something essential out of complexity. Forecasting alone takes just one line out of the given situation and follows the thread along, making no connections with the developments taking place at the time. The reciprocal elements and the reflexivity within the whole system are lost if forecasting is used as a one-and-only process. Therefore, a holistic strategy method would include both processes.

René Rohrbeck illustrates this necessary collaboration in a very understandable way. I met Rohrbeck in the cafeteria of the Deutsche Telekom tower, 60m high with a breathtaking view of Berlin. As you might have guessed, it's an inspiring surrounding for such a topic of discussion. He showed me his scientific classification of the strategic foresight processes from the article *Strategic Foresight* in *Multinational Enterprises – a case study on the Deutsche Telekom Laboratories*, written together with Heinrich Arnold and Jörg Heuer (sketched by Anna Luise Sulimma on the right side).

"In the 1970s, research conducted on the subject was termed 'forecasting' and focused on methods for predicting the future with modeling and econometric techniques, mainly using data from the past. These methods

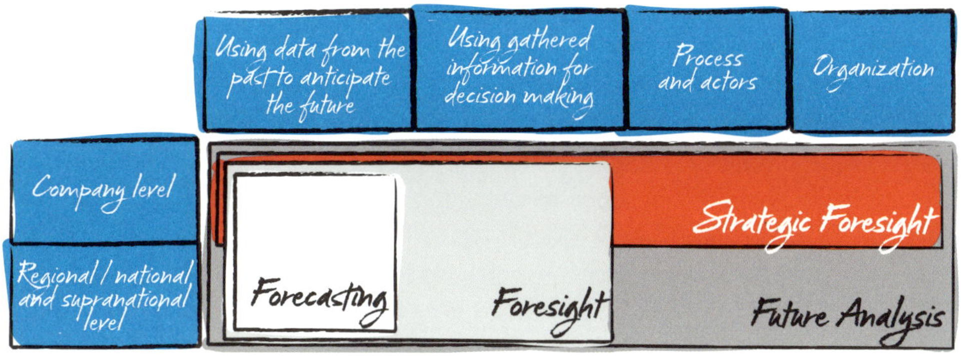

included trend extrapolation, S-curves, trend curves, and patent and publication analysis. Foresight broadened the scope of research to incorporate methods that enable networking for information gathering, assessment and interpretation, and methods that support decision making. Furthermore, foresight includes research on the capacity of organizations to cope with the future. In the 1990s, the scope of research was broadened by including the organization and the processes of the future investigation. The term 'strategic foresight' was developed to refer to research focused on the company level. **Today, to a large extent, 'future analysis' has substituted 'strategic foresight' as the preferred term.** In this article the authors use the term Strategic Foresight to emphasize the focus on the company level only".
At the same time, the use of foresight for companies can also be encountered with the name of 'corporate foresight' and has the same meaning. This term is becoming more professional and widespread, according to Z_punkt's management, a German foresight company, and it is not only used in strategy work but also increasingly connected to innovation development, marketing and R&D.

Hopefully now, the terminologies should be pretty clear. The topic of strategic foresight is so broad that it acts like the word 'trends' for the futures industry. There are a lot of touch points with other processes, methods and tools.

While speaking in Zürich with Adrian Müller from Noocleus, about where this process starts, I found out that he first tries to make customers forget who they are and what they do. He wants to make them think about society and how it will look 20 years from now. He says that he doesn't formulate specific pictures of the future because "foresight work is not so much about creativity but about imagination…you should be able to think about something that doesn't exist yet".

"I exactly knew the point to start my company, Noocleus. It was in my studies when my mentor mentioned 'strategic early warning' for the first time. I was intrinsically motivated to find out how companies deal with their external environment and how they integrate this knowledge in their strategic work. To say that I saw a need on the market to start such a service would be a nice answer but in fact, I was just fascinated by the subject. Companies usually know a lot about their market environment but not so much about the general environment. Foresight is about the general environment", says Müller.

Successful strategies are rarely achieved by spontaneous flashes of genius, but rather result from the systematic collection, analysis, and evaluation of facts, circumstances, trends, and opinions." / UK Government Cabinet Office – *Survival Guide*

But what is really necessary for the foresight process to be applied in a company? What does it entail? In short, as understood from Müller, strategic foresight requires a solid understanding of the future corporate environmental trends because organic growth means being consistent with future market requirements. Trend analysis, scenario techniques, wild cards, road mapping and all the other stuff we've learned about so far, represent a strategic foresight toolkit that enables the translation of the long-term observations into a well-founded basis for decisions. When you foresight, you think about and develop futures.

For UNITY, foresight is the instrument for inserting effectiveness into innovation. "Deciding on the directions where to go and committing on the fields to innovative, the foresight process structures and focuses the idea generation on those fields that are interesting for the future", says Michael Herbst, Business Unit Manager at Unity.

Rohrbeck and H. G. Gemünden in *Corporate Foresight: Its Three Roles in Enhancing the Innovation Capacity of a Firm*, identified three roles that strategic foresight should play to maximize the innovation capacity of a firm from an empirical study of 19 case studies and 107 interviews:

- The strategist role explores new business fields, creating a vision for the company, offering strategic guidance.
- The initiator role increases the number of innovation concepts and ideas, identifying new customer needs, technologies, and product concepts of competitors.
- The opponent role challenges the state-of-the-art of current innovation projects, scanning for disruptions that could endanger current and future innovations.

"The study shows how corporate foresight performing the initiator role increases the ability of a firm to produce incremental innovation. This leads to the conclusion that the interplay of corporate foresight and innovation management activities can contribute to organizational ambidexterity", say the authors.

A study conducted for the European Commission has shown that the time horizon of stategic foresight activities differs greatly, ranging from two to five up to 20-30 years (Becker, 2002). Some of that spread can be attributed to the type of industry or business and can be made more comparable if standardization on product life cycles is made", states Rohrbeck and Gemünden in *Strategic Foresight in Multinational Enterprises: Building a Best-Practice Framework from Case Studies.*

Research
Technology Management, argues that "the desired characteristics of a foresighter include being: a lateral thinker, knowledgeable in the scouting field, respected inside the company, cross-disciplinary orientated and imaginative.
/ Wolff M. in *Scouting for Technology*.

Output of R. Rohrbeck and H. G. Gemünden in *Corporate Foresight: Its Three Roles in Enhancing the Innovation Capacity of a Firm.*

// @FUTUREMANAGEMENTGROUP

Pero Mićić

I think each of us has somebody to turn to when things in a project get stuck and someone they can count on to bring insights and power thoughts to a topic. 'That somebody' for me is Pero Mićić, internationally renowned as one of the leading experts in future management, and founder of the FutureManagementGroup (FMG). In 1991, at the age of 24, while he was still studying, Pero Mićić started to work freelance as a future management consultant and has since dedicated his entire professional life to this fascinating topic. As soon as Nils put me in contact with Mićić, I knew he would have the answers to all my questions. I went to meet him on a Saturday at his office space in Eltville, a little town that I previously mentioned, in the middle of nowhere near Frankfurt. I felt truly privileged that he took time from his weekend to study my matter. Mićić has the most holistic view and knowledge on future related topics of anyone I met during my research for this book. His story says it all.

Back in 1991, Mićić was fascinated by the idea of looking ahead and knowing things earlier than other people do. He found everything he read about 'strategic early warning systems' quite boring: just systems of figures and indicators. I can totally relate to that! So when he came across an article written in 1975 by H. Igor Ansoff entitled *Managing Surprise and Discontinuity*, he got in touch with what is now called 'the future research discipline'. Back then, the world of business and the world of futures research were having difficulties understanding each other. Mićić felt that there needed to be a bridge between practical strategic management and futures research. This bridge is now the mission of his company.

FMG is unique because of the Eltville model which consists of "a process model that describes five different perspectives on the future [the so-called 'five futures glasses'] as a sequence of mental steps [for example in a workshop] and a results model, which describes the results achieved in the process as 'thought objects'", says Mićić.

THE FIVE FUTURES GLASSES

According to FMG's own definition, the Eltville Model is a result of observation and phenomenological analysis of more than 800 workshops with management teams. It was developed by Mićić and is now being developed further by the FMG employees. It serves as an integrated 'cognitive map' for terms such as scenario, vision, trend, wild card, assumption etc, which are frequently used in different ways.

So, in order to build innovation with a holistic approach, we need to wear all the glasses and ask ourselves the following questions:

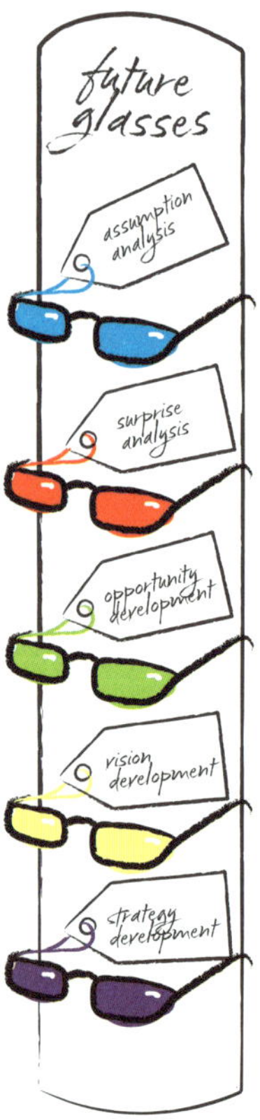

BLUE FUTURES GLASSES (assumption analysis): How will our market, work and life environments change in the next five to ten years? And what will remain the same?

RED FUTURES GLASSES (surprise analysis): How could the future surprise us and how should we prepare for possible surprising events and developments?

GREEN FUTURES GLASSES (opportunity development): Which threats and opportunities for new markets, products, strategies, processes and structures will arise from these changes and what can we do?

YELLOW FUTURES GLASSES (vision development): What does our company need to look like in five to ten years time in the sense of a strategic vision?

VIOLET FUTURES GLASSES (strategy development): How do we need to design our strategy to realize the strategic vision?

It has been experienced that the above sequence of wearing the colored glasses is also the most useful one but actually, there are 120 ways to combine the five futures glasses and each combination represents a new way to build innovation. There are some principles from which the choice of the sequence and combination should be derived but I will let you discover these on your own in Mićić's recently published book, *The Five Futures Glasses*. I didn't expect to hear such a complex issue in easy language. My favorite is the analogy between the futures glasses and the captain at sea which I will briefly show here in some excerpts.

THE WINDJAMMER CAPTAIN'S FIVE FUTURES GLASSES

"Imagine the captain of a tall ship, who sets out with his crew to distant, unknown destinations. In order to steer his ship towards a good future, he needs to regularly get a picture of how the sea and the weather will develop over the coming hours and days. The captain and his crew are dependent on natural forces. They cannot influence them, let alone steer them in a certain direction. Thinking creatively about the future weather is of no help to them either; fantasy is out of place here. The captain needs to get hold of available weather data and process it into his own personal assumptions about the future weather in the coming hours and days using his experience, powers of observation and logic. A clever captain will not rely solely on his own estimates. He will ask his officers and weather specialists for their personal assumptions about the future. He will not fully understand it and his crew, in turn, will not necessarily understand every aspect of his assumptions. A common picture of the future weather situation will only be arrived at gradually after discussions. A sail strategy can be developed based on these common assumptions about the future although there may be a certain variation, on assumptions about the wind direction for example. Ideally, it should also be possible to implement the sail strategy even if the deviating assumptions about the future later prove to have been more correct.

What color does the captain see when he looks at the sea and the sky? Blue! He is wearing the blue futures glasses for the probable future. The captain knows that only one thing is more or less certain on his journey into the future: that the future will surprise him..."

Refer to the book to find out what happens when the captain wears the other color glasses...

According to Mićić, upper management would benefit from this approach at the time of making decisions. "This model gives clear terms, a mental map and a real semantic network of terms and of objects of thoughts by connecting a general process and a general model of results. This is unique. We address to the top management team not to marketing, innovation people and strategic planners and this is different because many other approaches simply cannot be used within the top management team".

WEARING THE FIVE FUTURES GLASSES

There are different stages of maturity of clients, some of them know deeply what they need and some of them don't. FMG sees itself as a navigator and a synthesizer of future knowledge. In order to be safe and really informed when building a future strategy, companies must listen to all relevant sources: collecting all the trends, technologies and factors that futurists around the world publish and talk about. Then, FMG implements a semantic futures database called the 'FutureNet', like a knowledge network, that is tailored to the client. That's how it all starts; roughly, these three steps follow:

1. The client's needs analysis: identify, understand and try to help them understand what they really want. Identifying the right questions that need to be further answered.
2. The plan: develop a solution that is a project plan, a methods plan with which we can answer the questions.
3. The five futures glasses: mainly, this means identifying the factors, looking at their implications with the primary goal of identifying new future markets as opportunities and as threats. Future markets can be threatening because they could possibly replace what a company does.

I had woken up very early that morning, in order to catch the flight to Frankfurt and be on time for my meeting with Mićić. After two hours of intense discussions with him, while the coffees kept coming and going at a frenetic pace, my brain was wobbling over the notion of 'threatening' and even 'destroying' future markets. I'd never heard of these terms before. Mićić gives me the Gillette razor example; Photonics might lead to a threatening future market for Gillette. "The science of photonics includes the generation, emission, transmission, modulation, signal processing, switching, amplification, detection and sensing of light", says Wikipedia. Did you write that down? Because there's going to be a quiz later.

Through photonics we mean everything that is connected with lights, LEDs, lasers and connecting this field with shaving can lead to the creation of a laser razor. Why not? This would be a completely different product that could replace Gillette's razor function. In such a case, FMG would identify the potential of this new market, researching the existence of a company called Palomar that produces a laser razor (or its technology). A threat can literally be transformed into an opportunity if Gillette subsequently develops or buys this technology. "This is what happens in the first phase: we clearly identify and bring the assumptions of the management team to the surface, we challenge these assumption with our knowledge, perception, arguments, to make them robust and then we identify potential strategic surprises [like the one with the laser razor], a technological surprise", says Mićić.

// USING TRENDS IN FORESIGHT

FMG is continually looking at around 500 trends, technologies and issues that they call 'future factors'. It is all managed in a knowledge network, a semantic database. In our understanding, the clusters gathered are megatrends. Mićić explains, "The term 'megatrends' is a bit problematic in the way that people tend to reduce their thinking about the future to 10-12 megatrends. A few very boring megatrends – because all the people in the world know them. Which is ok for a start but if you want to go deeper and really find innovation and opportunity for innovations, than you need to be more detailed [working with microtrends and weak signals] and you need to change the focus from what is relevant to what can be relevant. If you really want to do foresight, then you need to establish new connections". Gillette would have never thought of photonics if they were not forced to, methodologically brought to do so. Another client of FMG who produces film (also called foil – something that covers buses with advertising, for example) would have never thought of displays if they weren't forced to. But they need to think about displays because displays are going to replace their business by providing the same function as the film but with a lot more advantages.

Together with their client, FMG decides upon the set of relevant future factors for a specific project by looking at the established questions and trying to make connections, mental links to the future factors. They usually end up with more than 100 future factors relevant for the company.

// WHAT IS FUTURE MANAGEMENT?

Seems like to foresight and forecast is not all you have to do to complete the innovation management process. But what else is there? At the beginning of this book I was wondering what the touch point is between innovation management, forecasting and foresight. I learned that foresight is a holistic process but just one of the wheels on the cart. With forecasting hopping in, it's still not enough.

Innovation management refers to the green glasses (aka opportunity development). "Foresight is literally looking ahead so I would use it to identify trends, make assumptions, spot surprises and opportunities. But foresight does not cover the entire process because it doesn't develop vision and foresight does not develop strategy", says Mićić. It covers the first three glasses but it doesn't cover the 4th and the 5th. So what is step four and five? Vision and strategy, it is future strategy.

Future management is the link between basic business strategy

processes and information related to the future, bridging these two at a top management level. "We are not talking about managing something literally in the future, but about people's current thoughts, assumptions, ideas, convictions, beliefs, hopes and fears. That is what makes the FMG approach to future management so practical and grounded on solid research at the same time", further clarifies Mićić. Future management helps CEOs take decisions, develop visions and grasp the essence of the future knowledge. Mićić explains, "Future management closes the comprehension gap between this abstract research, trend research and the practical requirements of managers in companies". I was never good at mathematics but this time I needed to add and subtract:

Future management is the bridge between strategic management on the one side and futures research on the other side. It describes the totality of all systems, processes, methods and tools for the early recognition and analysis of future developments and their inclusion in strategies. / Pero Mićić

FUTURE RADAR (THE FUTURE FACTORS RESEARCH)
\+ THE FIVE FUTURE GLASSES (FORESIGHT + FUTURE STRATEGY)
\+ INSTITUTIONALIZATION (MAKING FUTURE MANAGEMENT A REGULAR PROCESS)
= FUTURE MANAGEMENT

In the same brilliant book, *The Five Future Glasses*, Mićić states that: "We started using the term 'future management' in the mid-1990s when professionally managed companies increasingly and systematically concerned themselves with analyses of the future beyond strategic planning". Hmm, I'm confused. So, I asked the question I post whenever I don't really understand the process, but don't want to show it – I make it look like I'm just interested in the details: which is the input and the output of these processes we are talking about?
Mićić continues, almost ignoring my question like he gets it that I'm lost, "Managers and entrepreneurs often consider futures research and the related trend research as imprecise, noncommittal and unreliable. There is often a huge gap between their need for knowledge and the knowledge offered by futures and trend researchers. Future management closes this gap."
Futures research, which is often abstract and too theoretical – doesn't ask 'how to climb the mountain in a better way' but decides 'which mountains are best to climb'. Future management makes it easier.

IN THE FUTURE

"In the future, innovation management and future management which is kind of following will become mainstream, a necessity. There are a lot of approaches out there and no consultancy except for book keepers and tax advisors who do not talk about innovation. So, it's a very ubiquitous industry", says Mićić.

He still believes that innovation needs a guideline, a frame, and that the power of innovation needs to be directed within a company. Without knowing where the company is about to go and about the necessary improvements for the future, there is no real way to have effective innovation management. Due to the lack of the aforementioned, there are companies starting big innovation management projects that will have no impact on the market. In this understanding, FMG uses future management to control and direct innovation, identify the questions where innovation would be needed most at the moment and solve the bottle neck effect. It's a huge business and it's growing every day. More and more companies are entering the future business today and even the smaller ones have departments or assigned people that support these functions even if they are not called 'future something' or 'innovation something'.

Tomorrow I'm going to Hamburg and meet Marc Schüling to talk about strategic planning. It's been snowing continuously for 2 weeks here so I hope the trains will work properly. How time flies: it's already winter again. I feel like a little kid with all the snow, but still wish that somebody – anybody – would come and drag me (on a sleigh) back to summer.

identifying possible future states
grasping something essential out of complexity
strategic foresight
select the areas that should be examined in more detail via forecasting methods
searching for the areas you have to look at more in depth
one step in the planning and preparation of decisions concerning the future

STRATEGIC PLANNING

During my investigation, I found various definitions and words of advice regarding how strategic planning works. Some of it sounded like 'just grab a few key people from your organization, lock yourselves in a meeting room for hours and start planning strategically.' I find this approach doesn't fit with our fast moving world of today. Sitting down and doing visualization exercises regarding where you, as a brand, will be five years from now or only conducting a SWOT analysis (strength, weaknesses, opportunities and threats) probably won't do the trick either. Most of the tools and methods we have talked about so far can be easily integrated into strategic planning in order to make it relevant. Today, this process needs the input of forecasting, foresight, trend research and analysis. Strategic planning determines where the company stands in the current market and finds ways of getting you where you want to go as a brand, by helping managers make decisions and allocate resources.

There are 4 directions, formed in my head based on previous talks and reads, that became my 'question marks' over strategic planning's current meaning:

> **Strategic planning** in the sense of strategic management was left behind at the end of the last century. Its method was: we will think extensively about the status of our company and then we will develop a plan. It lacks many things, the most important among them being implementation and strategic change.

> **The act of strategic planning** traditionally means doing a lot of number crunching and analysis. The input is a concrete amendment coming from the top management related to how the market share of the company shifted in the last year and directions of what to do in order to shift it again.

> **Strategic planning** in advertising companies means a bit of consumer research, a bit of online research (in the ideal case this transforms into a trend research) and then developing a plan according to the results: what are we going to do with the brand (advertising ideas), with which

tools (aka channels) and for what purpose (who's our target group?).

> **Strategic planning** is a part of the innovation management process, a transition step from future research to its implementation. It tells you 'how' and 'when' to activate each step of your vision.

Marc Schüling

I tend to believe in and am more interested in the last two points of view. Who can help me solve the mystery of its true meaning(s)? I believe Marc Schüling can, and this is why I'm in Hamburg today. Schüling is a Strategic Brand Coach & Creative Visionary, an independent consultant since 2009 at Always-In-Beta. Getting in touch with him was a difficult task. After I sent him an absurd amount of emails, he called me one evening while I was attending my German course (spending 30 minutes with my mobile dictionary in front of the shampoo shelf trying to decipher which product was suitable for my type of hair led me to the conclusion that it was time to make my life more efficient and learn German). I called him back after my class but my phone crashed! To make things worse, after I got him back on the line and was just about to state my point and surely convince him to grant me an interview...my phone restarts itself! Alas, technology fails you when you need it most. Anyway, we managed to set up a meeting for the next week in Hamburg. Until then, I had time to prepare and devour any information I could find about strategic planning. Let's see...I know you are already familiar with the terms but I will repeat, once more, the core meaning of the three processes we've been dealing with in this book. Maybe you don't need this recap, but I do!

> **Foresight** is about considering different possible outcomes of the future.

> **Forecasting** is about predicting one future.

> **Strategic planning** is the overall set of methods and tools for making decisions that are relevant for the long-term.

My experience while reading future literature is that these terms can be very confusing. For example, scenario planning is close, as a word structure only, to strategic planning but they are two different approaches. **Scenario planning poses 'What if...?' questions that involve uncertainties outside the organization, looks at several alternative versions of the future (possible futures). Strategic planning refers to a single future, the so called preferred future.** Why the *preferred* future? Because the outcome of the process is a plan of how to achieve your visions brought forth by processes like forecasting and strategic foresight. As it is about planning and managing your objectives, it will always take the shape of something that you would *prefer* to happen to enhance the company's future.

According to Wikipedia, "Strategic planning and decision processes should end with objectives and a roadmap of ways to achieve those objectives". I will continue by paraphrasing now; simply put, strategic planning decides where an organization is going in the future and how it's going to get there. Setting goals and making a plan according to SWOT (Strengths, Weaknesses, Opportunities and Threats) analysis, PEST (Political, Economic, Social and Technological) analysis and working with data from the past, not taking into consideration trends or future market developments is what strategic planning used to mean and what some companies still continue to do. This older interpretation of the process related to the systematic programming of pre-identified strategies, is suitable for stable environments but not responsive enough for rapidly changing ones – it does not take into account unexpected events. The drawbacks of the process started being acknowledged in more and more working areas, causing some companies' to integrate scenario planning into their current process. **As this book is focusing on future thinking we will talk about strategic planning that uses information about the future as input.**

In case you are wondering why Wikipedia is such a popular source in my citations, it's because the topic that we're researching here is too new to be found offline. Thanks to the knowledge economy that we live in, we no longer have to go to libraries and dive our noses into books (even if this is still something enjoyable) because we can find out what we're looking for from the safety and comfort of our desks.

TRADITIONAL STRATEGIC PLANNING

Traditional strategic planning, according to the Millennium Project, consists of: "a mission, what the organization is all about; a set of goals, what it has to accomplish, generally in the relatively short and continuing run, to achieve that mission; and the objectives under each of those goals in order to move it towards practical applications. Finally, traditional strategic planning moves to detailed action plans". In accordance with this point of view, Pero Mićić declares that strategic planning is a subset of strategic management; it is the planning part. "If we look at the term planning, it means what we are going to do, when, who's going to do it and what are the resources for that. So, I would see strategic planning always as a sub-term of strategic management". But what has changed regarding this approach?

NEW STRATEGIC PLANNING

What about the terminology in itself? Michel Godet, in *A Toolbox for Scenario Planning* chapter published in *Futures Research Methodology* of the Millennium Project, talks about the abusive use of the term 'strategic': "How then does one find one's way around planning and strategic management? I remember Igor Ansoff telling me in 1986, when discussing the choice of a title for my book in English (Godet, 1987): 'you and I are well aware that it's the same thing, but sales will be better with strategic management'". In the same study, he also talks about the new strategic planning approach: "The last two decades have seen the popularity of strategic planning through scenarios soar, especially among large corporations in the energy sector, e.g. Shell and Elf. Since the early 80s we have sought to develop a high degree of potential synergy between prospective [futures] and strategy. The resulting synthesis was an integrated approach: strategic planning using scenarios". After further investigation I realized that more than being just a subset of strategic management, strategic planning is actually synonymous with it in most cases.

Michael Herbst from Unity declares his wish of making companies aware of this *new* way of conducting strategic planning by including information about the future in the process: "Most of the companies don't look at information about the future; they just look at where the competitors are, what the relative market share is, what the customers want today and then they make a plan out of it. Future foresight is a small part of strategic planning today. But we want to change that. In order to make the strategic planning process complete we need to include more future research and analysis tools".

Kerstin Cuhls, scientific project manager at the Fraunhofer Institute for Systems and Innovation Research in Germany, supports the new approach and mentions in our talk that, nowadays, 40% of companies integrate future focused methodologies into their strategic planning process, such as scenarios, Delphi and future workshops. The number is higher than I expected. When I asked her if there is a certain order in which these tools and processes are applied she comments that this is not necessarily the case: "At the Fraunhofer Institute we try to combine the methods in different ways. It always depends on the objective you have and in most of the cases you have the search (bibliometrics and internet) but then you can go directly for scenarios [foresight] or if you can cluster the information on your own you can even go for a Delphi and then build a roadmap. Forecasting starts when Delphi is applied or at the point when you have identified something at which you want to look at more in depth. Normally, after the Delphi assessments you look in depth [surveys] at what is really relevant and that is the moment when forecasting can be applied again. But on the other side, you can also start with forecasting old things and then put them into a Delphi. If you have alternative scenarios you can take the factors and forecast them once more to look at how scenarios will change if something will be completely different than at the beginning. It's like a loop".

Bibliometrics is a set of methods used to study or measure texts and information.

// @ALWAYS-IN-BETA

After stuffing all this information into my head, I met Schüling in Hamburg. He worked for a number of advertising companies doing strategic planning and he gives the perfect example and testimony to demonstrate the approach is suffering some changes. "During my work as a strategic planner at MEIRÉ UND MEIRÉ in 1997, I realized I am less interested in writing timetables and doing cost forecasts. I had a specific talent which was spotted and pushed forward by the creative director there – that was to write concepts for brands by inserting future research".

I was keen on finding out which of the four directions that I proposed on the first page of this chapter Schüling related to more. He tells me the story of strategic planning in Germany starting in the late 1990s: "The term itself didn't change or develop. The point is that all the agencies always had a strategic thinking oriented person sitting at their desks. It wasn't just a department or a job description but a talent. The projects got more and more complex over the time and it got difficult to be aware of all the elements that made up a part of the client's project. There was a need for a specialization that would insert into classical strategic planning work, spotting trends on the market and deep diving into consumer minds because the world became more complex. The German market found it necessary to implement this third dimension of the discipline into the agency structure".

Schüling calls the old approach of strategic planning, (in a diplomatic way), "an operating kind of planning" and he explains that it involves number crunching, market place analysis, market shares, doing plans and recommendations so company A beats company B. "This is not very strategic but rather tactical, which is not totally bad but strategic planning, in my view, is a visionary approach that tries to trigger things that are already there and push or invite clients to join a reachable future journey. Looking five to ten years ahead is not so far off or difficult for people to follow. Thus, above all, I try to enrich brands with visions". But is the vision time frame related to the type of industry? "The time frame depends on the task. Either way, strategic planning has to look into the future otherwise it is not strategic anymore. There are markets that are highly volatile like fashion: it would be difficult to forecast more than one year in advance and then there are other markets like atomic reactor brands that are very long-term oriented. Strategic planning has to think ahead of the current status quo", says Schüling.

It sounded to me like a big part of strategic planning was scenario planning. So, I posed the question being almost sure of the answer I would get: "**Scenario planning is just one tool within a giant tool box for strategic planners**. You can also use limbic maps [try to identify the motivational spectrum of your target group comparing your users versus the ones that reject your brand], or create question pools for 'benefit laddering' [trying to identify the emotional sphere on which your brand delivers and at what level your clients are capable of following you]. Or you can accept the rather easier and empirical side: that strategic planning's tools are a piece of paper, a pen and your brain". I know it's always hard to answer these questions because it is almost impossible to translate your mental process in words but I have to ask them. Thus, we have to understand that there are some artists and creative people in general who plan their work intuitively and others who use a more systematic method of planning their masterpieces. Such as Warhol, who called his company 'the factory' and had an industrial approach to creating art. Schüling is part of the first kind of planner with a rather intuitive approach: "Still, the outcome doesn't have to be just beautiful but also convincing, showing that you are informed, you analyze the market, know the clients and understand the competition. The way you do it…there are many variants".

Laddering is a qualitative marketing research technique, which seeks to understand why people buy and use products and services.

Schüling doubts market research. He is looking into the past *only* to understand the evolution of phenomena. He glimpses through all the data to find a connection so, he wouldn't neglect forecasting. The world is turning so fast that when you finish with your research and have the results, the world has gone and changed again – this is why Schüling's company is called Always-in-Beta. To prove his point, Schüling tells me about Nike in England: "they did around 12 projects per year without any market research. From 12, six failed, three were real successes and the rest were ok. In the meantime, other companies did three projects: one was a success, one a failure and one was ok but they went through a highly complex and very expensive research process to get to there".

As you probably have already understood, there is no unique or systematic way of doing strategic planning. It depends on the status of the market and your tasks as a brand. What is certain, as Schüling says, is that first you need to get into your target group's mind – sometimes even identify unknown target groups. How do you do this? It's a combination of trends and consumer research plus the continuous preparation of a planner: desk research, web research, listening to TED conferences and reading various coolhunter.coms (generalization). So are you ready then to start developing a strategy? "A planner is already developing his strategy while he is researching. It is important to constantly fuel your brain with ideas and perspectives on certain things. This becomes your mental patchwork that enables you to come up with new interesting combinations of information. Everything is already in front of your eyes; planners have certain capabilities to extract this relevant information", says Schüling. There are so many options out there that, I imagine, as a brand, it is hard to choose where to go. Companies need a guide, someone to say let's use option A, B and C for the next ten years and combine them in a new option D. This is a job for people like Schüling who focus on down to earth communication strategies, developing key messages, creating relevancy for target groups, and finding the right medium to do this. The kind of tasks a strategic planner deals with varies; Schüling is often focused on efficiency and tactics, but then there are clients who ask him to help them develop new products and identify new markets where their brand can jump in. I worked in an advertising agency for three years so I kind of know the game there. It's a lot of work and a lot of fun at the same time. I would say advertisers' creativity is in a way framed by the typical output of such an industry. What I mean is that when you contract the assistance of a full-service advertising agency you know that you will get either a TV ad or a PR campaign, or an online campaign together with some media space. Sure, there are loads of ways to do this in an innovative way. But still, creatives often have even greater ideas that could exist outside the sphere of these boxes. These ideas are sometimes new business opportunities, products or even new services. This is great because the way we communicate and perceive the things around us has changed and the shift to being involved in a variety of communication channels (at the same time) has been made. The problem is that creatives don't really know what to do with these ideas. They are obliged to deliver ads and texts. I believe more and more advertising agencies have detected a gap. Schüling, coming from the advertising world, found the solution and created a trend based innovation company that overcomes this problem: "We are still at the very beginning with the project. After generating the ideas, when it comes to the operational part, things often collapse so that's why Shakers & Makers was founded, to help these ideas see daylight". Does all this mean that advertising's business model will slightly adapt or that we are witnessing the rise of a new kind of service? I'll keep an eye on the evolution and perhaps write about it in a second edition of this book.

John Naisbitt said that "Strategic planning is worthless, unless there is first a strategic vision". As I see things, there are various ways of building this strategic vision. I have shown here a few tools and processes to do this and now it is up to you to decide which way to go. If I were you, I would try to work with most of them in order to obtain the holistic approach.

Can we build a holistic map out of all these elements we've discovered and talked about so far? I will do so in the next chapter.

On an end note, one of my favorite authors and foresighters, Kerstin Cuhls, expounds on the future of these three processes: "the evolution of semantics will develop and change the way we use these processes and we will learn how to better combine them. We will all become more pragmatic and the methods and tools will especially embed strategic planning at the end of the road".

We are close to the last stop – the end of my road trip dear readers, but I bet you are wondering what the future brings for the innovation industry. Actually, nobody knows for sure but there are a few people who at least have a vision. I met some of these experts on a quick trip to London and at LaFutura – remember the event where futurists and trendspotters from all over the world gather in Berlin for one day, to exchange insights? Let's hear from some of these insightful people again.

the preferred future
future
future
future
future
strategic planning
the overall set of methods and tools for making decisions
that are relevant for the long-term
decides where an organization is going in the future
and how it's going to get there

A NEW UNDERSTANDING OF INNOVATION?

The Future Navigator

THE FUTURE NAVIGATOR

Recently I had the pleasure of being invited to the TEDx/frog event about Beauty/Complexity in London, so I took advantage of this alluring city and met with two companies that I have admired for a long time: ?WHAT IF!, the world's largest independent innovation company and The Futures Company, the coming together of Henley Centre HeadlightVision and Yankelovich – the grandfathers of future studies. My thoughts on innovation were in their final stages but the road trip pressed on.

First I was off to meet Dave Allan, co-founder of ?WHAT IF! who explained to me that back in 1992, when ?WHAT IF! was founded, "innovation was important but not necessary or urgent. Today, everything we do needs innovation and it became a fundamental commercial discipline that tends to be called 'innovation management'. Innovation is not about being lucky but about a replicable management discipline for your business, a central discipline and not just 'nice to have' anymore". Why now? Well, because companies don't have all the tools in-house (e.g. the production lines) anymore but they mainly work in partnership and open innovation. You think Apple produces any of its products or components entirely by itself?

For Allan, "Innovation is the process and delivery of new things that make the world a better place. If let's say you need to 'explore' to get oil, innovation equals the exploration process in itself – it's about finding new opportunities, starting off new things and inspiring people to imagine new possibilities".

With a view of the Tower Bridge, just outside The Futures Company office, I was pondering my talk with Allan and getting ready for my next inspiring meeting with Will Calgey, Global CEO for The Futures Company. There is something about these companies and being in London that makes you feel you are at the center of the world's interest. "**It is very easy to be interesting when you talk about the future.** Being useful by choosing to work with the right tools is harder. Today, innovation is too intangible and doesn't have clear values attached but we're trying to get beyond this buzz word and find out what's the problem you need to solve. It surely isn't only about new product developments but about unlocking growth by redefining your competitive market. Innovation is like a lens to the future but it relates more to understanding the past and present rather than the future", says Calgey. I am reminded again about the importance of history and present events in shaping future developments. The Futures Company emphasizes this thinking and is the only company that has an actual timeline of their extensive past on their website.

London opened my appetite for more interpretations of the word 'innovation'. My colleague at TrendONE, Sebastian Metzner, who I secretly nicknamed 'the Brain' (he just informed me that he earned his Super Mayor badge on Foursquare but has realized how unimportant this is for his real life), was sitting next to me so I paused to ask him what 'innovation' means for him. It turns out Sebastian believes that innovation behaves like nature and therefore is the result of a self-regulating system that lives in harmony, due to its perfectly matching elements. Because of their high complexity, these kinds of systems cannot be managed. Does this mean that innovation cannot be managed?

He explains. "As humans we think we can control everything but we can't. Imagine innovation like a natural ecosystem. What we can manage are its elements but not the interaction between the components. Therefore, good innovation is the result of good environmental conditions: people, ideas, relationships, tools, creative surroundings etc. Nature cannot be completely controlled (see biodiversity issues), nor can innovation. Openness and lack of structure are our drawbacks and our advantages at the same time".

I can totally relate to his interpretation. Innovation is like culture. Can you control culture? It has infinite forms and possibilities and is fluid, in a constant state of flux.

// WATCH THINK INTERPRET ACT

For quite some time now, Nils and I have been talking to people about how the innovation industry needs a framework and a common language. Do you remember the story I told about consultancy companies in the 1960s and how organized they have become nowadays? Well, we wish the same thing would happen to the trends and innovation disciplines. Would this be beneficial? Sure it would, if we want to grow we need to find the proper directions, but can this industry be confined to some rules and predefined outcomes? We're not sure about that yet.

Questions abound. What should innovation have as an output? Can we define the output of such a complex process that needs to be tailored according to every project? And if we could, what would this look like? Is control not the answer?

I still don't know the answers to all of the aforementioned questions, but I believe the first step towards identifying them is to put everything on the table. On the same canvas that is. Therefore, I am grouping the topics that we have learned about during our road trip, into impulses, methods, tools and processes. They all help us grasp innovation and by putting them in a singular visual representation we can see how they interrelate. Of course they are all different, of course some of them can merge at a certain point but they have one thing in common: they all claim to enable innovation. Therefore they must be on the same page, right?

When I felt lost and alone during this road trip my friends and mentors helped me get back on track. Especially Torsten Rehder, TrendONE Knowledge Director, and a most appreciated supporter and power booster. He was always there when I needed him, coming up with coherent opinion and good critique. Among many of his generosities was the invitation to lecture at the Macromedia University for Media and Communication in Hamburg where he was teaching a Trend Research module for the Media Management Bachelor degree. For one hour, I talked to the students about the contents of this book: five minutes about each topic. It was the first time I heard myself speaking aloud about the methods, tools and approaches and it was extremely useful in helping me realize what was missing or what was unnecessary. The audience was a big asset as they were asking the right questions that turned into targeted answers in some of the chapters you have now read.

Keeping everything together, in spite of the complexity and the differences of the topics, is possible because the elements can either be implemented individually or form part of an innovation toolbox. How you choose to combine them depends on each project. Complexity is exactly what makes them usable in many situations. The urge to find a relationship between every part lead me to creating this visual, my brainchild born from this process, this love affair.

It is the Future Navigator map...

The Future Navigator

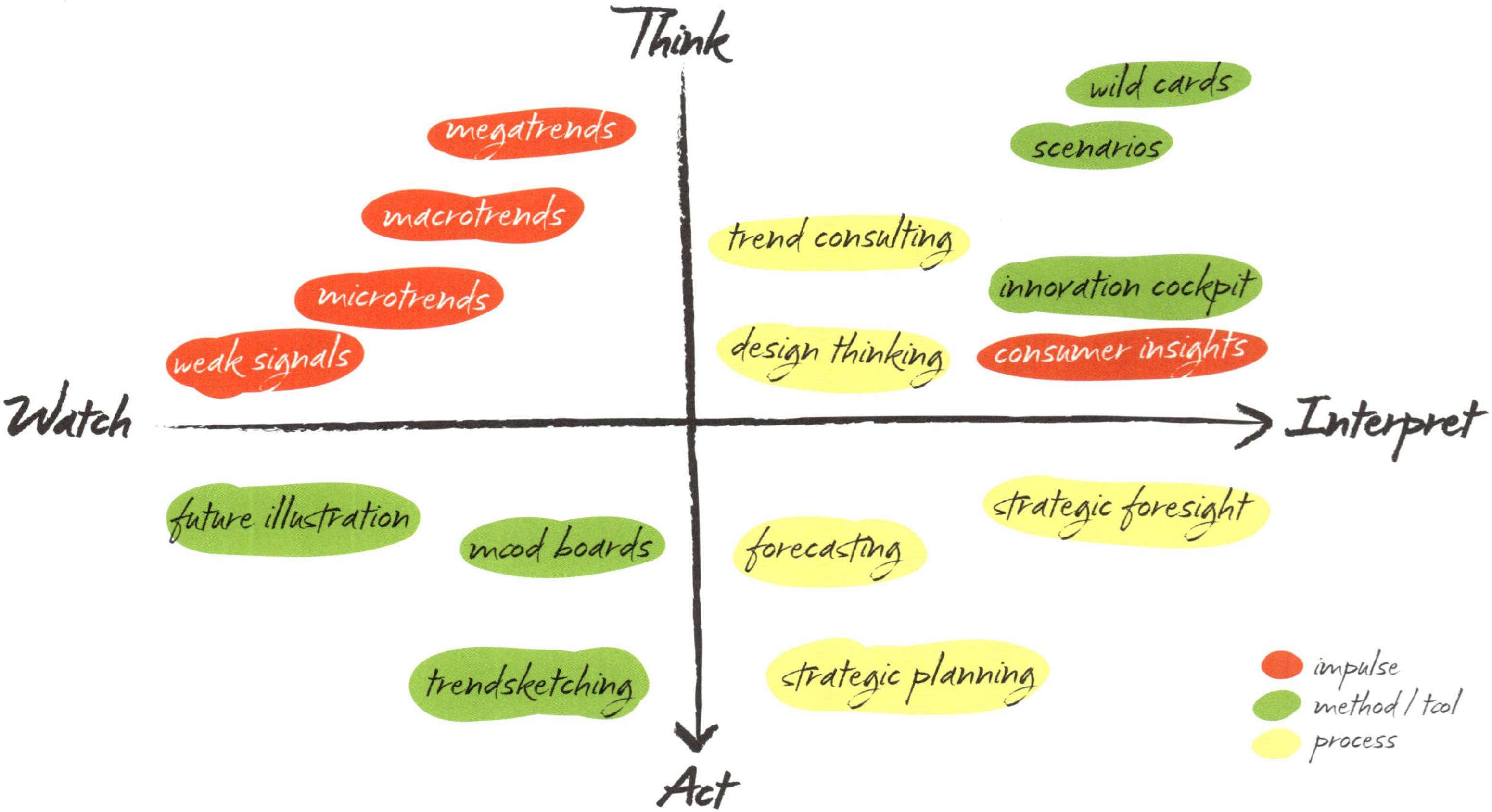

Tada! The idea of the Future Navigator acting as a framework is to design a space where everybody can find their place. In order to be able to do that, I needed to extend the area as much as possible, because the spectrum of the activities is so large that it made it difficult to find just one box for all the approaches to fit together.

The Future Navigator is a generic map not an innovation model. It explains and presents methods, tools and approaches by showing their common fields. It is not supposed to be followed step by step in a sequence, but to inform you of the disciplines for spotting the future that are available for you to select in the next innovation project.

As you may have already realized, none of the topics that we've met during our road trip exclude one another. They co-exist and moreover, they complement each other. Could we then build a framework for the innovation industry starting from these approaches? Is it too simple to explain the innovation world in such easy steps? Well, it is a beginning – just as Naisbitt pointed out when he explained the whole world through ten megatrends.

If you have read the pages of this book and didn't start at the end, you must know that I've been trying to keep things as simple and clear as possible all along and that's exactly what I did with this map as well. I drew two axes and assigned values: think-act (from low involvement with the client to high involvement) and watch-interpret (from the market to the consumer).

- > **Watch** is about researching, looking for signals and trends.
- > **Think** of their meaning for today's society, their interdependence and impact in general.
- > **Interpret** refers to extracting meanings and insights from a research process and eventually applying it for building a new brand or service strategy. Defining the information in a context.
- > **Act** is the implementation, the concrete steps that have already been defined, through various research methods: who, when, where, how.

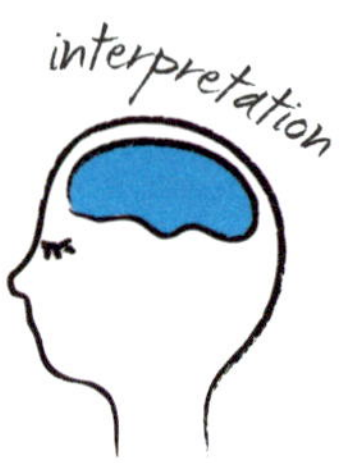

The axis from 'Think' to 'Act' refers to how involved the agency is in the process – from thinking and just giving ideas to actually helping to make them happen. 'Watch' refers to the market whereas 'Interpret' refers to the consumer dimension. The difference between 'Think' and 'Interpret' lies essentially in the path from only thinking and detecting the phenomenon that surrounds us in itself to actually interpreting what they mean for a specific client context. Furthermore, to 'Act' is about how to fulfill all of our wishes after thinking and interpreting them. Of course, everything in itself involves thinking and interpreting, but we should try to keep these terms only in these frames for now. (If you start analyzing, everything can be a method or a tool for something and anything in itself can be or build a process).

A process is a regular sequence of tasks and activities to achieve a goal or results whereas a signal is a piece of information about possible developments and events in the future.

If you follow the map from left to right you will notice that it starts with impulses and ends with processes and goes from a short-term to long-term look into the future. In an innovation project, the axes don't need to be followed in any one particular sequence. Of course, it is natural to start with detecting the information and then reflecting on it, but visualization tools for example, help you and can come in whenever you need them: at the very beginning or more at the end of the project. The most complex part is the last quadrant on the right side – applying – when it comes to complex processes that involve most of the previously mentioned tools and methods. And now that we have them all together on one page, it should become easier to understand their essence and relation – even though processes, tools and methods as categories are mixed. But as we know, they are mixed in real life too!

Impulses are often integrated into methods. Methods use tools, and processes use all of these. Alternatively, each could be a standalone activity.

// LA FUTURA

Does the trends and innovation industry wish for a framework and set standards? LaFutura proves that the answer to this question is yes. LaFutura was a unique event that represented a great start in the building of an innovation community. Usually, people from this industry never have the chance to meet. But on November 11th 2010, 70 leading futurists, trendspotters and innovation experts from all over the world, gathered for one day to talk about their industry in the PanAm Lounge and President's Suite in Berlin. This is where Kennedy used to stay – an amazing place where time seems to stand still. Everything from the furniture to the PanAm flight attendants uniforms remained intact in their 1960s state of perfection. Great minds attended the event, which enabled an excellent network exchange. The attendants forgot the word 'competition' and shared their thoughts and discussed key future insights.

Besides creating a network of futurists and trendspotters from all over the world, LaFutura's vision is to bring more understanding, reflect on and define the standards of excellence for the trends and innovation industry. Everything started with a vision and some questions regarding a deeper understanding of our existence in this business sector:

> “Are we just a handful of lively, small entrepreneurs with a unique insight into the future? Are we a new breed of market researchers? Are we corporate consultants specializing in the latest trends & innovations? Are we more of a patchwork of insight, foresight, mega-, macro- and microtrends, scenarios and wild cards? Wouldn't it be good to be considered as part of a new business sector? I genuinely believe that the best way for us to arrive at a deeper understanding of our true nature is through intelligent co-opetition, a more transparent overall system and greater harmonization of our vocabulary. These conditions are necessary for our customers to perceive us as a business sector and for enabling us to do our jobs as successfully as possible.
> / Nils Müller, Founder & CEO of TrendONE ”

Some professionals that I talked to over the phone before the event were skeptical about this gathering and co-opetition idea. They wanted to know why they should participate and share knowledge that they spent a lifetime creating and gathering. Nobody at LaFutura actually asked them to divulge their secret models (though open innovation is about growing together by sharing our knowledge) but they didn't want to partake anyway. That's when I wondered about three things:

1. How these experts got to be so highly appreciated in the future industry without spreading what they have learned.
2. Where the industry is heading if they prefer to die with their knowledge as opposed to passing it along and
3. Are they just a band of information freeloaders?

I strongly believe the future stands for cooperation and open innovation. In a world of widely distributed knowledge, LaFutura's goal is to break boundaries and go beyond competition, as joint insights are the key to progress in any field.

LaFutura wants to grow the industry by setting up a framework and create a network of inspirational people. Besides getting to know each other, the event proposed 5 working topics and let the attendees choose what they want to rack their brains with: *Methods of the trend and future landscape, Creating a TrendUniverse, Picture* (as in visualize) *the future, What material is the future made of?* and *How do innovations come into existence?* If you guessed I was attending topic #1 you are correct! Myself and ten (+/-) trends and innovation experts, many of which we have heard from in this book, were assigned the room with a fireplace. As 'flight attendants' served us hot tea and coffee, we received craft materials, gathered around tables and put on our thinking caps. The output of our brainstorming had to be a work of art – I love the fact that we had to use our hands to actually build our thoughts. Our spontaneous masterpiece sort of took the form of a construction site, which seemed appropriate due to the topic.

I took advantage of my time there and asked these experts what they thought of a holistic toolbox for building innovation and about my Future Navigator. Actually people ended up trying to build an innovation model out of it, which was not its initial purpose but hey, no rules, no boundaries! We all started brainstorming: should we have loops and circles instead of an axis? As innovation is a looping and dynamic process should the rings be: scanning, translating, building a concept or a vision, communicating it, making the decision, executing it and then starting all over again? "Our branch is based on the idea of change. We need a system that scans the environment and makes the customers understand that innovation is an ongoing exercise", contributed Michael Herbst, Business Unit Manager at UNITY. Then, should the Future Navigator look more like a spiral?

In spite of the fact that innovation may not be possible to structure, Herbst thinks we need to build up systems that make the change manageable and operational for the dimensions of daily business: R&D, marketing etc. There are so many parties involved in this game and they all have to be aligned to the same mindset.
Poetically, Herbst also believes that we should actually start with labeling things: "We tame wild horses but actually what we have to work on right now is to put the labels on the cages in the zoo. It is difficult to build a business because the clients don't understand what the difference between our companies is. If we could find together a framework where it is easier for each of us to explain what we can do and what we can't do, this would help our industry grow".
I said this at the beginning as well, most of the companies I met on my road trip, cover more than the topic I describe but they do things at different 'flight levels'. Some of them work at a microtrend level some at mega or some start creating a product that will hit the market in three years. They are different businesses and have different targets to meet. Cornelia Daheim, Managing Director at Z_punkt, says: "The four quadrants in the Future Navigator show that you have different entrances into the universe of trends and innovation, different targets on the horizon and different tools to use accordingly".

One day, futures research may become an organized body of assumptions and methods with a more formal academic tradition; in the meantime, it can be thought of as an art in that it is creative and/or as a craft in that it applies knowledge with skill. / Jerome C. Glenn in *Futures Research Methodology Version 3.0* from The Millennium Project.

Florian Peter from CScout, also commented on the simplicity of the map: "It should be so simple that everybody can use it as a starting point to explain our industry. When we approach companies, if we have a general overview on the trend and futures industry, it is then our decision to take the project to 5000 or just to 1000 feet. If we successfully position ourselves into these quadrants, then we can grow as an industry and explain ourselves better". In conclusion, we should look at the Future Navigator as an organic and adaptable structure.

Herbst dots the 'i' by stating: "We should find a way to put all our assets together and explain to people what we do because there is no competition at the moment. The decision makers don't know why they should choose one method over the other and that's because the industry was not able to articulate it so far".

Regarding what level you address in the company or what the needs are, Daheim adds, "You can apply many of these methodologies for different needs and you can scale them up and down, using them and sorting them in various ways. Are you taking them as far as making a decision or even implementing it or not? And for that you can use trends or scenarios and add Delphi and modeling and many more. The methodology as such can be applied in different ways, depths but it depends on what kind of need you support. That's one way to sort the methods better".

All of these thoughts on the status of the trends and innovation industry coming from professionals, some whom have been dealing with the topic for decades, gave me a nice feeling that I was on the right track in trying to solve a relevant difficulty of the industry: tidy up the methods.

So, people from the industry, are you ready to define your category? Ten years ago partners of the same industry would not accept meeting and talking freely about their businesses. Now they do. The popularity of events like LaFutura and the open participation in this book from experts in the fields of trends and innovations may in fact mean that we are indeed ready.

// THE INNOVATION ALLIANCE

A few innovators in this field have already started to define their purpose within the industry and focus on what they excel at. These companies realized that their approaches match together with other company's approaches and form an all-inclusive innovation model that now activates under the name of the Innovation Alliance (IA).

The main idea behind the IA is that companies should not limit their services to their working model anymore but simply add others to create a holistic approach to building innovation. The future stands in using complementary methods and instruments.

The Innovation Alliance was born in TrendONE's kitchen in Hamburg, after Nils Müller and Sven Tollmien worked together with Sturm und Drang on a large project. They suddenly realized that the working process should be somehow institutionalized so both the clients and the partners involved can profit from it. Therefore, the IA is actually the starting point of a framework for the innovation industry. "Moreover," says Nils, "trends and futurists are seen as freaks working separately. We want to build up industry logic, create a structure and standardize our industry so it can grow. Otherwise it will stay like a freak show. The alliance is not only about spotting the future and trends but also about processing and realizing the projects together with innovation experts, doing the implementation and translation of the research".

At LaFutura we had the chance to see the quantity and variety of approaches that exist out there and everybody was surprised by the interdisciplinarity of the field: from trendsketchers to hard-core consultancy, everything is present, a colorful crowd of approaches. This is great, but it makes the client dizzy; he needs orientation as well as the industry. "Within IA we found out how to work together. At first, it was very difficult but the IA is close to finalizing a framework, a holistic innovation model", says Nils. More about this is soon to come.

IA's objective is to build up an industry logic that structures the various approaches of trend, future and innovation in order to enable growth. Building standards to enable growth.

IA is open to companies from all over the world who wish to join. Its criteria being a unique USP (unique selling proposition) that fits in the working model – a smart strategic differentiation. The alliance doesn't have barriers because everybody goes in with its core competence, so even though they are

competitors in a way, they can work together and complement one another. If companies come together and work in an open innovation model does this mean that innovation has changed? IA says nowadays we are dealing with a new type of innovation: agile innovation – it comes very fast, the life cycles of products get shorter and shorter so there is a need for a new school of thinking, a new innovation approach. "Theoretically, this means you can't conduct a project in a process anymore but you should work in a sort of spiral-like way, always bouncing back up and changing things. You need to be very fast and flexible", says Nils.

Not to throw around the word 'love' but I love Nils' inability to explain something without sketching it on a piece of paper:

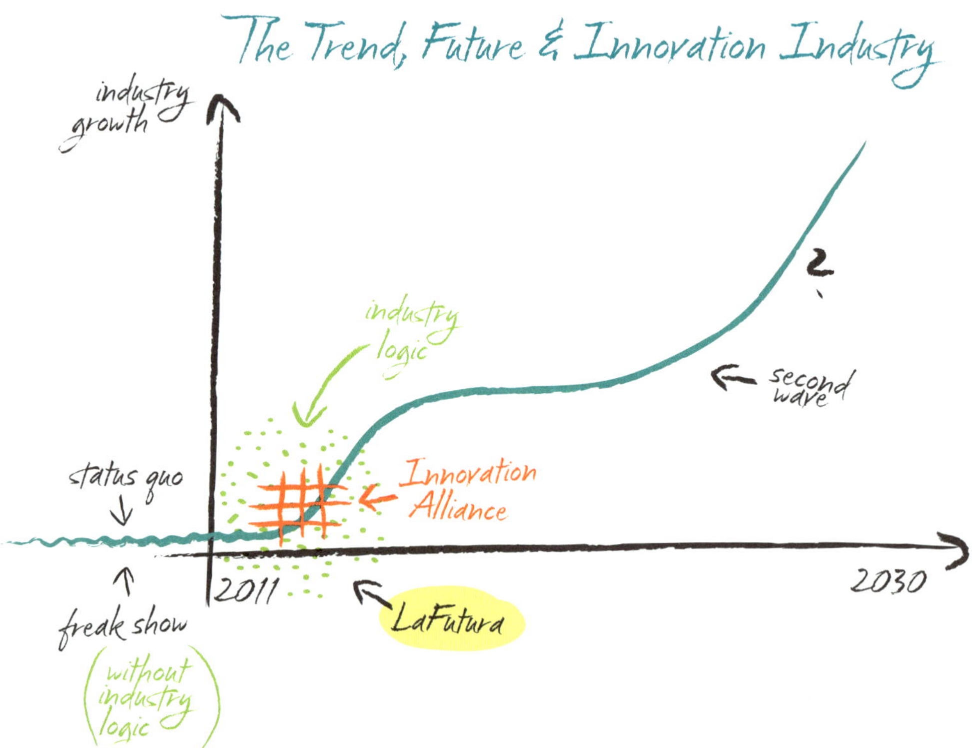

// A NEW NATURE OF INNOVATION

As my curiosity was awakened earlier at LaFutura regarding open innovation as a redefined way of seeing productivity and growth of a business, I looked for more experienced opinions. Z_punkt states in the article *Trends – Innovation Communication. Dialogue Replaces Knowledge Transfer* by Klaus Burmeister and Silke Loh, that "Open Innovation is an important trend today: The research and development department no longer has a monopoly on devising novel products, processes, and services. Innovation does not originate anymore solely from companies concentrating know-how internally, developing new ideas, and making them ready for the market. Co-innovations, such as the joint development of a hybrid motor by Mercedes and BMW, are becoming more frequent. The marketing is then carried out separate".
I would allow myself to state that the peak of open innovation in the publishing industry might have been made by the Forbes 'crowdsourced' December issue in 2010, which stated, "The package is entitled 'Names You Need to Know in 2011' and I need your help to pull it off". They wanted to know from the community who are the people, places, products and ideas that will have significant impact over the coming 12 months. It was a thrilling experiment that worked out perfectly: the Forbes community had the chance to put whatever content they wanted in the magazine and Forbes looked 'cool' because they let them. To enrich their open innovation experience, some companies actually build platforms (sort of closed topics related social networks) and work with them every day creating new products, evaluating prototypes and starting all sorts of discussions. My Starbucks Idea asks customers and interested parties to post, vote on, and discuss ideas and improvement suggestions. Swisscom, as I read in *Trends – Innovation Communication. Dialogue Replaces Knowledge Transfer*, asked its lead users: "How can cell phones be used for the authorization of particular services? Out of 44 ideas, six were developed further". From the same source, I read about IDEO, which also demonstrates a well thought out example of an open innovation platform, called 'OpenIDEO'. It can be used collaboratively by all interested members to create and develop ideas and design processes. "The purpose of this platform is to solve tasks and problems

through crowdsourcing: the process begins with a problem or challenge introduced by IDEO. Over various steps, users can then send in their suggestions and solutions, compare them with those entered by other members and then decide whether or not to develop them further. The first project worked on in this way was the design of the platform itself".

The New Nature of Innovation report to the Organization for Economic Co-operation and Development (OECD) was jointly conducted by the Danish and Finnish governments' analyses on new emerging forms and areas of innovation and how governments are responding to a 'new nature of innovation'. My eyes popped out of their sockets when I read this the first time and it still has a big impact on me. At the very beginning of this road trip there was a hint of a new nature of innovation and now towards the end, our understanding of what this exactly means is coming together: "A new nature of innovation is emerging due to a transformed business environment". The innovation economy, the globalized world, the individual put in the center of everything, these are all things that contribute to a new nature of innovation. At this point we still have to ask ourselves, what exactly is so new about it? "Technology has always played an important role in driving innovation, and it will continue to do so in the future, but for many companies technology will gradually move from being a driver of innovation to becoming an enabler of innovation. In the future, companies will need to become more open; i.e. to learn from their customers, collaborate with others [aka open innovation] and assume greater social responsibility. If they fail to do so, they will not survive. While companies will still optimize their business, globalization and the digital technology have changed the rules of the game", say the authors in the same report to OECD.

According to the *The New Nature of Innovation* report, there are four drivers that will tailor the future of innovation:

1. Co-creating value with customers and tapping knowledge about users
2. Global knowledge sourcing and collaborative networks
3. Global challenges as a driver of innovation
4. Public sector challenges as a driver of innovation

I recommend a further read of the study to find out about other principles that will change our innovation behavior in the future. In the meantime, I bring you Verena Kuen to reinforce the aforementioned statements...

THE BUSINESS SHIFT FROM R&D TO MARKETING INNOVATION

To build innovation no longer means creating a 'breakthrough' or working with an R&D department, innovation might be achieved simply by giving play to a good idea coming from (perhaps) a microtrend you spotted.
"I had a dream the other day. I was in a time machine, brought back to the 1970s and I said 'Yeah! I know everything because I saw how the future looks and I can become the first trend researcher ever'. All of a sudden I realized, 'where is my laptop? I don't have a cell phone! Where's Google? How can I reach out to the world?' And then I woke up! Innovation is strongly related to the availability of information". Meet Verena Kuen, a freelance innovation consultant from Vienna. We were in a very French café in the district of Prenzlauer Berg (where everyone in Berlin goes to breed and eat cakes) having an enjoyable talk about innovation in big organizations. Kuen started working in pharmaceutical marketing 11 years ago, after a long career in consumer goods marketing. This past year I have only seen the agency's point of view on innovation so I was anxious to hear her insider's opinion in regards to how innovation is seen within companies. I asked a few basic questions:

How do you define innovation? "For me, innovation is about placing things in a new context and creating additional value for the client.
You don't have to develop things from scratch. Creating a new packaging that is biodegradable for example, in the pharmaceutical industry, is a huge step forward".

How does innovation translate in a pharmaceutical company? "It used to be mainly about R&D activities like creating new clinical entities, new drugs but now, along with the business shift, innovation is translated also into new approaches for marketing activities and services around the brand, packaging developments and the implementation of new relevant ideas".

Verena Kuen

How do you get inspired? "We heavily work with consumer trends, benchmarks, we show success stories to our clients, organize expert workshops and do trend studies on target groups".

What part of the innovation process do you find most difficult to work with? "The actual implementation of something new is the hardest part. In big organizations, you need to create complex internal processes to make things happen. For example, if we do social media, the organization has to create social media guidelines and policies before. Also, a tough part is to convince people to work in a creative way every day. In many big organizations, working creatively is just the fun part, the add-on, the things you do only when you have time. They are just not used to it. So, we do small steps to make them feel comfortable in working with new ideas and innovations".

Why is innovation such a buzz word? Did you notice any shift in businesses' culture related to innovation? "Because of the availability of the information since the Internet exploded, five to seven years ago. Big organizations only do incremental innovation and they are less daring. With the saturation of the markets and the cost cuts, they realized that this concept is not working anymore. The topic of marketing innovation came up with the move to digital because we need a new type of marketing".

What about the next big shift? "Probably when the digital natives will be predominant in the work force. This is something that will really change the game for big companies. They are hard workers but very focused on work-life balance. The concept of work time and leisure time is obsolete; we're running away from the concept of waking up in the morning to go to our office and come back in the evening just in time for our leisure time.

We are not going to work less in the future but working time and leisure time will melt".

Well then, you'd better have a job that you enjoy because it is going to affect your whole life! Then, I suddenly made the connection with something I read from Z_punkt: "In 2015 co-creation in open information ecosystems will be the predominant paradigm. High performance businesses understand the needs of the millennial generation [aka digital natives] and provide solutions that help their future workforce to maximize productivity". Watch out, they're coming soon!

WHAT'S TO COME?

Richard Watson, futurist writer and consultant, in his "Extinction timeline 2000 – 2050," predicts the end of futurists. Giselle Weiss, journalist at Global investor 2.09 Focus_04 asked him why he thinks the future will have no need for trend researchers: "My theory is that in 50 years' time or whenever, the world will have settled down and become much more predictable. And we will also possess a phenomenal amount of data, thanks to technology. So I think futurists will just sort of die out. We should put on our deep-thinking caps, collectively agree on a future that we like the sound of, and then invent it. To some extent, we can do that. Thinking is a powerful tool". In accordance to this *thinking* line, Ray Kurzweil, an inventor and futurist that we could also call a future-life-tech guru, states in his thrilling book *The Singularity is Near* that our evolution is exponential and our way of thinking is still linear. Therefore, our mental structure is not built to imagine a future that brings, for example, the fusion of human and artificial intelligence. But this doesn't mean that if we can't imagine it, it won't happen. Kurzweil predicts this will happen sooner rather than later.

In the context of technological evolution I continued on the same line with Max Celko, a trend analyst & innovation consultant. During a phone conversation, he told me about the 'meta brain' a sort of scary system for sharing feelings and experiences:

Ray Kurzweil has been described as "the restless genius" by the Wall Street Journal. A leading inventor, he has been awarded the MIT Lemelson Prize, one of the world's biggest prizes for innovation.

"I am entering the coffee shop and letting my eyes sweep across the place. Text messages from my webcommunity commenting on everything I see appear instantly before my eyes – they're commenting on the patrons' looks and their manners of dress. I join their discussion through a direct link between my brain and the Internet. In a corner of the coffee shop, a person starts flashing red: a young woman I got to know last weekend. As I have downloaded all of her perceptions from the past 25 years in the meantime, her life is as familiar to me as my own. That is how I know when and with whom she exchanged her first kiss – and what that felt like to her. This scenario is enabled by brain scanners that read our sensory impressions, thoughts and feelings, and broadcast them to the net. Augmented reality contact lenses project virtual additional information directly into our fields of vision and so enlarge our perceptions of our environments.
/ Max Celko, *The Meta Brain – Merging Into The Human Cluster*"

The paper is a play of thought on what our future might look like by extrapolating trends that we're currently seeing and it shouldn't be understood as a clear prediction. Apart from providing food for our brains, Celko is mostly active in technological trend research and consulting. He states that: "Technological innovations are being developed and brought to market in increasingly short intervals, and the accompanying changes in society are becoming ever more global in reach. For this reason, companies are now obliged to recognize the potential of new technologies and shifts in consumer behavior as early as possible, and translate this potential into their own market offerings. Trend and future research is thus becoming increasingly important for identifying new business opportunities, acting as a central driving force for innovation."

Regardless of the media communication process and everything else becoming faster and faster, the speed of a trend may not necessarily change. But for sure it will affect *something* in the probability of *something* becoming a trend. In the future people will not receive the same information because media will be individualized – as seen in Celko's 'meta brain' as well. This probably means that we will deal with an eclectic trend context and it will probably be much harder for a fad to grow.

The most important skill in our rapidly changing society isn't to be able to add to the knowledge of a particular subject, but rather to be able to throw old knowledge aside and look at the world with fresh eyes.
/ Professor Feiwel Kupferberg from the Danish University of Education

FROM TRENDS TO INNOVATION AS AN ATTITUDE

After a year of asking people what innovation is, I think I've made up my mind and have built a point of view of my own. Whatever path you take to get there, it's important to understand that **innovation is about disruption, change and chaos**. Therefore, you need to be ready to stir up your brand. To achieve innovation, you need time to build up a mindset and turn on the future thinking button of everyone around you. It doesn't just happen when you snap your fingers because innovation is something organic and it's something that you build a relationship with, hopefully a long-term one.

I can't help but notice that innovation projects are somehow very personal and subjective. Your ideas depend on your experiences, your likes and dislikes; the creators' brains somehow tailor the projects. I experienced the best ideas coming from merging poly-social thinking material. **The richness of mixed elements always results in creativity, experimentation and change**.

Remember that you always have to feed innovation in order to keep it alive: trends, doing cross-thinking and unexpected combinations of information by keeping your eyes 360° wide open 24/7. Creativity is not confined to gallery or concert walls – it should infiltrate every part of our daily lives. And my last, but certainly not least, observance is that you need the right partners to help you in the process.

Tom Kelley in *The Ten Faces of Innovation* reminds us to, "go for distance. Innovation is not just a program; it's a way of life-just as the ultimate wellness program is not a fad diet but a healthy lifestyle".

Together with my fantastic Knowledge Unit colleagues at TrendONE, I have learned that everyone around us is a potential knowledge carrier. We do not live in an ivory tower! There is always something new to know. Disseminating the information is also an important point in *raising* the knowledge level. Starting conversations is the best way to solve the *mysteries*. From my experience, the most interesting insights and connections come up when you spread the knowledge so people can join in with additional information, thus elevating you to a new point. I hope I did bring you to a new knowledge point along this road trip.

I would have to write a few more books to include all that I have learned during this year, such as how to ask the right questions, to listen and sometimes to insist, to sit down quietly and intensively think for a moment, to dig for the meaning of things to bring them to surface. And of course, I learned what innovation is truly made of: knowledge, disruptive ideas, chaos, boldness, people's guts, a strong attitude and an open mind.

All year long I've been telling everybody that this book is due by Christmas, well it's Christmas and so my work is done for now. I am back home in Bucharest, surrounded by my family who is not interested in the future in the same way as the people I met during my road trip. This is unexpectedly and *damn* relaxing. All the future related topics mentioned by my grandma (for sure she has some '*scenarios*' already built in her head) refer to whether I will get married and have children soon and whether my parents will finally renovate their home as they have planned.

So, I'll see you around!
I suppose I have to think about my grandma's future ideas for a bit.

ROAD TRIP PROPS

Finally

THE TREND & FUTURE DICTIONARY

A collection of definitions picked up from interviews, articles, books and blogs discovered and used during the road trip.

WEAK SIGNALS

News stories or observations about technological and social innovations, posts in the social media, observations of novelty products, or simply a modest wall-sticker of an alternative movement. Weak signals are current oddities, strange issues that are thought to be in key position in anticipating future changes in organizational environments. / E. Hiltunen

e.g. H&M starts selling vintage clothes.

MEGATRENDS

Last decades, affect many different aspects of society, and involve a complex process that often includes politics, economy and technology.

e.g. Sustainability

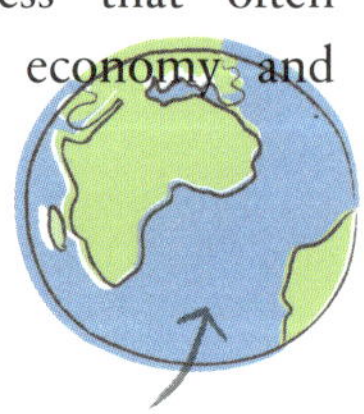

TREND

The general direction in which something tends to move.

MACROTRENDS

Pattern-based understanding of past and present, help to determine the likelihood of future events. Their life span is five to ten years.

e.g. Fair Trade

MICROTRENDS

Concrete examples of marketed inovations: technologies, products, start-ups etc.

Life span of 1-2 years until they develop into a stronger trend or disappear completely.

A microtrend is something new, intelligent, mass-market ready and structure changing.

e.g. Ben&Jerry's fair tweets: Ben & Jerry's and Twitter use every unused character by automatically adding a message linked to the tweet. The aim is promoting fair trade.

CONSUMER INSIGHTS

A fresh and not-yet obvious understanding of customer beliefs, values, habits, desires, motives, emotions or needs that can become the basis for a competitive advantage. / M. Sawhney.

Insight is about what's happening in the space between the consumer and the product, his key motives, drivers and barriers.

Spotted through close observation, empirical research and direct content with the consumer.

e.g. People startig to consume regional products being away from their hometwon. Insight: people are feeling home-sick.

DESIGN THINKING

A human-centered approach to innovation that draws from the designer's toolkit to integrate the needs of people, the possibilities of technology, and the requirements for business success. / D. Brown.

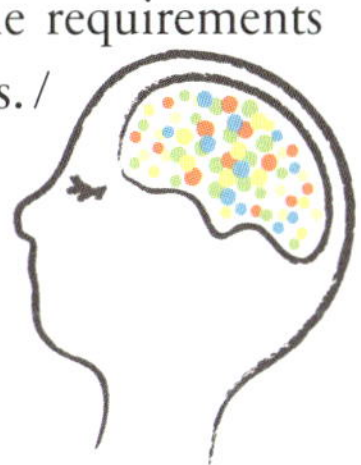

USER DRIVEN INNOVATION

Customers are observed to understand how they work with the product and what unrecognized needs they have. / Design Thinking is a broader approach and describes a whole process while User Driven Innovation just indicates that the user is integrated in your project development.

User-initiated innovation: the consumer creates or improves a product. Von Hippel called this 'user innovation'.

CREATIVE THINKING

The term innovation is often used to refer to the entire process by which an organization generates creative new ideas and converts them into novel, useful and viable commercial products, services, and business practices, while the term creativity is reserved to apply specifically to the generation of novel ideas by individuals or groups, as a necessary step within the innovation process. / Wikipedia

OPEN INNOVATION

A paradigm that assumes that firms can and should use external ideas as well as internal ideas, and internal and external paths to market, as the firms look to advance their technology. / Wikipedia.
Networked and collaborative innovation perspective.

FUTURE ILLUSTRATION

The use of illustration to trigger future thinking.

TECHNOLOGY ROADMAP

A plan that matches short-term and long-term goals with specific technology solutions to help meet those goals, applying to a new product or process, or to an emerging technology.
It provides a mechanism to help forecast technology developments. / Wikipedia

INNOVATION COCKPIT

Data based controlling instrument for innovation development; deals with trends and other technical, social and political influencing factors.

SCENARIOS

Projections that describe a complex picture of possible futures and possibly the path to them. Professional scenarios often rely on asking what changes in the present would be required to arrive at envisioned alternative future states. / Wikipedia

e.g. South Korea and North Korea will be united in 2028.

WILD CARDS

Low-probability, high-impact events. Such sudden and unique incidents might constitute turning points in the evolution of a certain trend or system.

e.g. In 2020 there will be no trace of oil on Earth.

STRATEGIC/CORPORATE FORESIGHT

Identifying possible future states (alternative future scenarios) for which sometimes no previous data is available.

Systematically thinking about and developing alternative futures.

Foresight is the product of insight.

Foresight marks the ability to see through the apparent confusion, to spot developments before they become trends, to see patterns before they fully emerge, and to grasp the relevant features of social currents that are likely to shape the direction of future events.

/ *Coping with the future: developing organizational foresightfulness*, H. Tsoukas, J. Shepherd

e.g. In 2050, the work force might drop/ decrease.

FORECASTING

Attempts to predict future states from current trends by making estimations.

Focused on methods for predicting the future with modeling and econometric techniques, mainly using data from the past.

Is about predicting one future (one scenario) from known current trends by making estimations and looking at quantitative research.

e.g. In 2050 80% of the population will be over 60 years old.

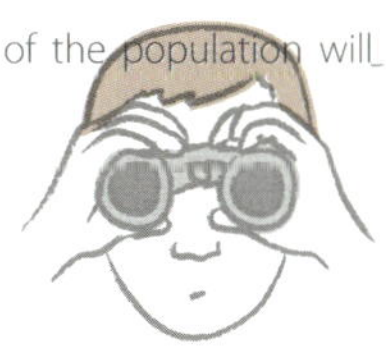

FUTURE MANAGEMENT

The bridge between futures research on the one side and strategic management on the other. It describes the totality of all systems, processes and methods for early perception and analysis of future developments and their inclusion in strategy. Systematically recognizing the future of markets and developing practical and implementable strategies out of these findings.

/ P. Mićić

FUTURE THINKING

A mindset. This is about how to deal constructively with the future as a person and as an organization.

TECHNOLOGY FORESIGHT/TECHNOLOGY INTELLIGENCE

Foresight applied in technology. Methods for collecting, analyzing and disseminating technological information have become known as Technological Forecasting, Technology Foresight or Technology Intelligence.
Technology Foresight further broadened the scope of research into the prediction of future technological developments.

e.g. For Deutsche Telekom: provision of relevant information on technology and the evaluation of their impact on the corporation.

STRATEGIC THINKING

Strategic thinking is intent driven. It provides a point of view about the long-term market or competitive position that an organization hopes to achieve over a defined time period. / *Competing for the Future*, G. Prahalahad and C.K. Hamel.

STRATEGIC PLANNING

Strategic planning deals with the preferred future and involves the overall set of methods and tools for taking decisions on how to reach these 'future states' that are identified as relevant through forecasting, foresight or other methods.
Develops operational plans for preferred futures with time horizons of one to three years.
Simply put, strategic planning determines where an organization is going over the next year or more and how it's going to get there.

e.g. To prevent the workforce from decreasing, we need to start creating job opportunities and assignments for people who are aged over 60. To achieve this goal, we need x, y, z.

ACKNOWLEDGEMENTS

At first it was just an idea and then, the road trip started, thanks to *Nils Müller* who knew I had the skills to do this before I did. Or maybe he just took a wild guess?! Thank you for giving me the chance to write this book and for pushing me forward every day. I am blessed to work with such a brilliant mind and visionary man.

Thank you TrendONE for commissioning the book and for your admirable patience in waiting for it to be completed.

Who else came with me on the road trip?

Thank you

Torsten Rehder — for being a power booster and such a positive influence, for coming up with good critique and for always being available to listen to and solve my worries.

Christina Bifano — for having the patience to fight with my text, for bringing your editing skills and knowledge to look for inaccuracies and inconsistencies, and for coming up with helpful suggestions.

Richard Skrein — for the super grammar check and enthusiasm

Orkide Ergün — for designing the book and bringing your lovely spirit into this project.

Jasmin Nauber — for layouting the book and making it nice and shiny. Additional thanks for bearing with the never ending corrections. How many beers do I owe you now?

Anna Luise Sulimma — for accompanying my text with your beautiful illustrations.

Bianca Stockreiter — for the encouragement, corrections and care.

Pero Mićić — for answering all my scientific questions and structuring my initial thoughts on innovation – showing me where to start from.

And of course, a big **THANK YOU** to all my interviewees, for their time, revealing thoughts and kindness in answering all my wild and sometimes naïve questions. One more thing, thanks to my parents for supporting me in going out in the world to meet extraordinary people and do great things.
Thank you all for the power, trust and motivation. It has been a wonderful journey with you besides me. Hats off!

SOURCES

Accenture (2010) *Information 2015: Reforming the Paradigm*. www.accenture.com

Aguilar, F. (1967) *Scanning the Business Environment*. London: Macmillan

Andresen, K.S., Kruse, M. Persson, H., Mogensen, K.Æ., Eriksen, T.T. developed by Copenhagen Institute for Futures Studies (2010) *10 Megatrends toward 2020*. www.cifs.dk

Ansoff, H.I. (1975) Managing Strategic Surprise by Response to Weak Signals. *California Management Review*, vol. XVIII no. 2

Ansoff, H.I. (1976) *Managing Surprise and Discontinuity: Strategic Response to Weak Signals*. Brussels: European Institute for Advanced Studies in Management

Avery, M.L., Kreit, B., Falcon, R. developed by The Institute for the Future (2009) *Food Web 2020*. www.iftf.org

Ballou, S., Chu, J., Morrison, G.P. developed by IBM Business Consulting Services (2004) *Deeper Customer Insight: Understanding Today's Complex Shoppers. www.ibm.com*

Bisson, P. (2010) *Why Trends Matter*. McKinsey Quarterly, www.mckinseyquarterly.com

Blackmore, S. (1998) Imitation and the Definition of a Meme. *Journal of Memetics – Evolutionary Models of Information Transmission*, 2

Brown, T. (2008) *Design Thinking*. Harvard Business Review, 6/2008

Burmeister, K., Loh, S. developed by Z_punkt (2009) *Trends – Innovation Communication: Dialogue Replaces Knowledge Transfer*. www.z-punkt.de

Celko, M. (2009) *Das Meta-Gehirn: Wie wir im Menschenschwarm aufgehen. / The Meta Brain – Merging Into The Human Cluster* Style and The Family Tunes Magazine

Copenhagen Institute for Futures (2009) *Future-driven Innovation*. www.cifs.dk

Costanzo, L.A. (2004) Strategic Foresight in a High-Speed Environment. *Futures*, 36/2

Cova, B., Kozinets, R., Shankar, A. (2007) *Consumer Tribes*. Oxford: Butterworth-Heinemann

Cuhls, K. (2003) *From Forecasting to Foresight Processes* – New Participative Foresight Activities in Germany. Journal of Forecasting, 22/2003

Cuhls, K., Ganz, W., Warnke, P. developed by Fraunhofer Institute for Systems and Innovation Research (2009) *Foresight Process on behalf of the German Federal Ministry of Education and Research: New Future Field*. www.isi.fraunhofer.de

Dalkey, N.C. developed by RAND Corporation (1969) *The Delphi Method: An Experimental Study of Group Opinion*. www.rand.org

Davies, R. (2008) *Shirky, Twitter, Sweeping Generalization*. www.russelldavies.typepad.com

Dawkins, R. (1976) *The Selfish Gene*. Oxford: Oxford University Press

Diamond, S. (2010) *Unconventional Wisdom*. The HUB, 22

Ferguson, S. (1993). Strategic Planning for Issue Management: The Communicator as Environmental Analyst. *Canadian Journal of Communications*, 18

Florida, R., Kenney, M. (1990) *The Breakthrough Illusion: Corporate America's Failure To Move From Innovation To Mass Production*. New York: Basic Books

Gilbert, D. (2006) *Stumbling on Happiness*. New York: Alfred A. Knopf

Gladwell, M. (March 1997) *The Coolhunt*. The New Yorker, www.newyorker.com

Gladwell, M. (2002) *The Tipping Point: How Little Things Can Make a Big Difference*. New York: Little Brown

Glenn, S. C., Gordon, T. J. (2009) *Futures Research Methodology Version 3.0.* The Millennium Project

Gloor, P., Cooper, S. (2007) *Coolhunting: Chasing Down the Next Big Thing.* New York: AMACOM

Godet, M., Monti, R., Meunier, F., Roubelat, F. (1999) *A Toolbox for Scenario Planning*. Futures Research Methodology Version 3.0

Godin, S. (2008) *Tribes: We Need You to Lead Us.* New York: Penguin

Grosskurth, J. developed by STT Netherlands – Study Centre for Technology Trends (2010) *Futures of Technology in Africa*. www.stt.nl/africanfutures

Grove, A. (1999) *Only the Paranoid Survive: How to Exploit the Crisis Points That Challenge Every Company*. New York: Doubleday

Guice, J. (1999) Designing the Future: the Culture of New Trends in Science and Technology. *Research Policy*, 28/1

Hafkesbrink, J., Evers, J. developed by Innowise (2011) *Innovation 3.0: Embedding into Community Knowledge. The Relevance of Trust as Enabling Factor for Collaborative Organizational Learning*. www.innowise.de

Hafkesbrink, J., Schroll, M. developed by Innowise Research & Consulting (2010) *Business Model Innovation in the Digital and New Media Economy*. www.innowise.aroew.spirito.de

Hajkowicz, S., Moody, J. developed by The Commonwealth Scientific and Industrial Research Organization (March 2010) *Our Future World: An Analysis of Global Trends, Shocks and Scenarios*. www.csiro.au

Hayward, P., Morrow, R. (2009) An Integral Approach to Scenarios. *Journal of Futures Studies*, 13/3

Hiltunen, E. (April 2008) The Future Sign and its Three Dimensions. *Futures*, 40/3

Hiltunen, E. (February 2009) Scenarios: Process and Outcome. *Journal of Future Studies*, 13/3

Hiltunen, E. (November 2006) Was It a Wild Card or Just Our Blindness to Gradual Change? *Journal of Future Studies*, 11/2

Hiltunen, E. (June 2010) *Weak Signals in Organizational Futures Learning*. Helsinki: Aalto University, school of Economics

Hiltunen, E. (2007) *Where Do Future-Oriented People Find Weak Signals?* FFRC-Publications, eBook 2/2007

Hofstede, G. (n.d.) *Uncertainty Avoidance Index*. www.geert-hofstede.com

Inayatullah, S. (February 2009) Questioning Scenarios. *Journal of Future Studies*, 13/3

Innovation Point (n.d.) *Customer Insight as a Driver of Strategic Innovation: Non-traditional Approaches for Breakthrough Customer-Inspired Innovation*. www.innovation-point.com

Kalbreier, L. (2010) In Search of Tomorrow's Global Megatrends. *Credit Suisse – Global Investor*, 2.09

Kelley, T., Littman, J. (2005) *The Ten Faces of Innovation: IDEO's Strategies for Defeating the Devil's Advocate and Driving Creativity Throughout Your Organization*. New York: Doubleday

Kurzweil, R. (2005) *The Singularity Is Near: When Humans Transcend Biology*. New York: Viking Penguin

Kuusi, O., Hiltunen, E. developed by Finland Futures Research Centre (2007) *The Signification Process of the Future Sign*. FFRC-Publications, eBook 4/2007

Larsen, G. (2006) Why Megatrends Matter. *Futureorientation*, 5/2006

Lindkvist, M. (2010) *Everything We Know is Wrong: The Trendspotter's Handbook.* New York: Marshall Cavendish

MacDonald, J. (2009) *The first 6 Macro Trends of the 21st Century.* www.21stcenturymacrotrends.com

Mannermaa, M. (2004) *Heikoista Signaaleista Vahva Tulevaisuus*. Helsinki: WSOY

Martino, J.P. (1975) *Technological Forecasting for Decision Making.* Amsterdam: Elsevier

McCandless, D. (2010) *Information Is Beautiful. London*: HarperCollins

McGonigal, J. (2010) *Gaming Can Make a Better World*. www.ted.com

McKeon, M. (2008) *The Truth About Innovation*. Harlow: Pearson Education

Mendonça, S., Pina e Cunha, M., Kaivo-oja, J., Ruff, F. (March 2004) Wild Cards, Weak Signals and Organisational Improvisation. *Futures*, 36/2

Mićić, P. (2006) *Das ZukunftsRadar: Die Wichtigsten Trends, Technologien und Themen für die Zukunft*. Offenbach: GABAL

Mićić, P. (2010) *The Five Futures Glasses: How to See and Understand More of the Future with the Eltville Model*. Basingstoke: Palgrave Macmillan
Mogensen, K.Æ. developed by Copenhagen Institute for Futures Studies (2006) *Creative Man: The Future Consumer, Employee and Citizen.* www.cifs.dk

Molitor, G.T.T. (2009) Scenarios: Worth the Effort? *Journal of Future Studies,* 13/3

Müller, A.W., Glesti, S. (2009) *Visual Futures: Visualisierungen als Instrument strategischer Foresightarbeit*. Züricher Hochschule der Künste

N+1 (2010) *What Was The Hipster? A Sociological Investigation*. New York: n+1 Foundation

Naisbitt, J. (1982) *Megatrends*. New York: Warner Books

Naisbitt, J. (2006) *Mind Set! Reset Your Thinking and See the Future*. New York: HarperBusiness

Organization for Economic Co-operation and Development (2009) *New Nature of Innovation*. www.newnatureofinnovation.org

Penn, M., Zalesne, E.(2007) Just 1%: The Power of Microtrends. *Change This*, 38

Penn, M., Zalesne, E. (2007) *Microtrends: The Small Forces behind Tomorrow's Big Changes*. New York: Twelve

Petersen, J. (1997) *Out of the Blue: Wild Cards and other Big Future Surprises: How to Anticipate and Respond to Profound Change*. Arlington: Arlington Istitute

Pfanner, E. (2006) *Agencies Look beyond Focus Groups to Spot Trends*. The New York Times, www.nytimes.com

PriceWaterhouseCoopers (January 2008) *How Consumer Conversation Will Transform Business*. www.pwc.com

Randall, D. (2011) The Foresight of Crowds. *Design Mind*, 13

Rehder, T., Hentschel, R. developed by TrendONE and Proximity Germany (2009) *The Outernet: Say Hello to the Wild World Web*!

Rehder, T., Metzner, S., Stockreiter, B., Puhe, O. (2010) *Trendbook 2012: Das Zukunftslexikon der wichtigsten Trendbegriffe*. Hamburg: TrendONE

Rogers, E. (1962) *Diffusion of Innovations*. New York: Free Press

Rohrbeck, R. (2010) Harnessing a Network of Experts for Competitive Advantage: Technology Scouting in the ICT Industry. *R&D Management*, 40/2

Rohrbeck, R., Arnold, H., Heuer, J. (2007) *Strategic Foresight in Multinational Enterprises: A Case Study on the Deutsche Telekom Laboratories*. www.rene-rohrbeck.de

Rohrbeck, R., Gemünden H.G. (2010) *Corporate Foresight: Towards a Maturity Model for the Future Orientation of a Firm*. Heidelberg: Physica-Verlag

Rohrbeck, R., Gemünden H.G. (2008) *Strategic Foresight in Multinational Enterprises: Building a Best-Practice Framework from Case Studies.*

Rohrbeck, R., Heuer, J., Arnold, H.M. (2006) *The Technology Radar – an Instrument of Technology Intelligence and Innovation Strategy.* www.rene-rohrbeck.de

Rohrbeck, R, Hölzle, K., Gemünden H.G. (2009) Opening up for Competitive Advantage – How Deutsche Telekom Creates an Open Innovation Ecosystem. *R&D Management*, 39/4

Saghafi, F., Aliahmadi, A., Azadnia, M. (2009) Trend Mutual Effects Analysis Methodology for Strategy Configuration and Foresight. *World Applied Sciences Journal*, 6/9

Schoemaker, P., Gunther, R. (2002) *Profiting from Uncertainty: Strategies for Succeeding No Matter What the Future Brings.* New York: Free Press

Sheate, W., Zamparutti, T., Bennett, S., Rogeli, M. developed by The European Environment Agency (2007) *EEA Research Foresight for Environment and Sustainability: Final Report.* www.eea.europa.eu

Shirky, C. (2009) *Here Comes Everybody: The Power of Organizing Without Organizations.* New York: Penguin

Silvan, S. (2006) *Valppaus on Valttia: Heikot Signaalit Löytyvät Läheltä.* Helsinki: Talentum

Smith, C. (2010) *Microtrends – Today's Business Paradigm.* GDS International, www.gdspublishing.com

Sperber, D. (2000) *An Objection to the Memetic Approach to Culture*. www.dan.sperber.fr

Stone, M., Bond, A., Foss, B. (2004) *Consumer Insight: How to Use Data and Market Research to Get Closer to Your Customer*. London: Kogan Page

Tsoukas, H., Shepherd, J. (March 2004) Coping with the Future: Developing Organizational Foresightfulness. *Futures*, 36/2

Van der Heijden, K. (March 2004) Can Internally Generated Futures Accelerate Organizational Learning? *Futures*, 36/2

Vejlgaard, H. (2007) *Anatomy of a Trend*. New York: McGraw-Hill

Verhaeghe, A., Van den Bergh, J., Schillewaert, N. developed by InSites Consulting (July 2009) *Teen-Memes: Memetic Branding and Identity of Youngsters*. www.insites.eu

Von Hippel, E. (1975) *The Dominant Role of Users in the Scientific Instrument Innovation Process*. MIT Libraries, www.dspace.mit.edu/

Weiss, G. (April 2010) We Can Actually Invent a Future We're Happy with. *Credit Suisse - Global Investor*, 2.09

Wilkinson, L. (1995) How to Build Scenarios: Planning for "long fuse, big bang" problems in an era of uncertainty. *Wired, Scenarios: The Future of the Future Special Issue*

Wise, E., Høgenhaven, C. developed by Nordic Innovation Centre (June 2008) *User-Driven Innovation: Context and Cases in the Nordic Region.* www.nordicinnovation.net

?What If! The Innovation Company, www.whatifinnovation.com
always in beta, www.always-in-beta.com
Context-Based Research Group, www.contextresearch.com
Copenhagen Institute for Future Studies (CIFS), www.cifs.dk
CScout Trend Consulting, www.cscout.com
D-School: HPI School of Design Thinking, www.hpi.uni-potsdam.de
Fraunhofer Institut für System und Innovationsforschung, www.isi.fraunhofer.de
frog, www.frogdesign.com
FutureManagementGroup, www.futuremanagmenetgroup.com
gravity, www.gravity-europe.net
Helsingin Sanomat, www.hs.fi
IDEO, www.ideo.com
KEY VALUES, www.key-values.com
Max Celko, www.maxcelko.com
noocleus, www.noocleus.ch
ORKIDEE, www.orkidee.de
Pantopicon, www.pantopicon.be
The Futures Company, www.thefuturescompany.com
The Institute for the Future, www.iftf.org
The Millennium Project, www.millennium-project.org
TrendONE, www.trendone.com
TrendSketcher, www.trendsketcher.de
trendwatching.com, www.trendwatching.com
UNITY, www.unity.de
What's Next, www.nowandnext.com
What's Next Consulting, www.whatsnext.fi

INDEX

// ABOUT THE AUTHOR

Proudly hailing from Bucharest, Romania, ***Delia Dumitrescu*** draws from a communication and public relations background. Allured by semiotics since the early days of university, she tailored her favourite study to trend research and analysis. While living in Italy and Spain she became deeply involved in visual communication and coolhunting and published several texts and works of photography related to consumer investigation. In the spring of 2010, while sharing and building knowledge at TrendONE Germany, she set-off on a road trip to find out what innovation is truly made of, and hasn't stopped since. Currently, Delia is living and seeking inspiration in Berlin.

// ABOUT TRENDONE

THE MIND IS LIKE A PARACHUTE. IT WON'T WORK IF IT ISN'T OPEN.

TrendONE, the leading microtrend research company in Europe takes pride in opening your mind – providing comprehensive reviews of the latest developments in the market and boosting the inspiration of your business and thinking. With a worldwide network of scouts, TrendONE looks for intelligent marketing campaigns, new technology, media and online innovations, new business models and products. As Nils Müller, Founder and CEO of TrendONE says, "The future will only come to those who believe in it!"
/ www.trendone.com

IMPRINT

Road Trip to Innovation – How I came to understand Future Thinking

Author: Delia Dumitrescu
Editor: Christina Bifano
Grammar Check: Richard Skrein
Art Direction: ORKIDEE
Layout: Jasmin Nauber
Trendsketches: Anna Luise Sulimma

Printed by: Grafisches Centrum Cuno GmbH und Co. KG

Printed in Germany

ISBN: 978-3-00-035736-7

TrendONE GmbH
Gasstraße 2
22761 Hamburg
Germany
Tel: +49 (0)40 526 778 0
info@trendone.com
www.trendone.com